The Art and Science of Writing a Scientific Paper

The Art and Science of Writing a Scientific Paper

Editors

Devendra Mishra MD ACME FIAM

Professor
Department of Pediatrics and
Department of Medical Education
Maulana Azad Medical College, New Delhi
drdmishra@gmail.com

Dheeraj Shah MD MNAMS FIAP

Professor
Department of Pediatrics
University College of Medical Sciences, New Delhi
shahdheeraj@hotmail.com

CBS Publishers & Distributors Pvt Ltd

New Delhi • Bengaluru • Chennai • Kochi • Kolkata • Lucknow • Mumbai
Hyderabad • Jharkhand • Nagpur • Patna • Pune • Uttarakhand

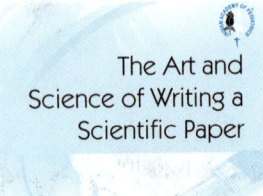

The Art and Science of Writing a Scientific Paper

ISBN: 978-93-89261-80-6

Copyright © Indian Pediatrics

First Edition: 2020

Reprint: 2023

Published by Satish Kumar Jain and produced by Varun Jain for

CBS Publishers & Distributors Pvt Ltd

4819/XI Prahlad Street, 24 Ansari Road, Daryaganj, New Delhi 110 002, India.
Ph: 011-23289259, 23266861, 23266867   Website: www.cbspd.com
Fax: 011-23243014   e-mail: delhi@cbspd.com

Corporate Office: 204 FIE, Industrial Area, Patparganj, Delhi 110 092
Ph: 011-4934 4934   Fax: 011-4934 4935   e-mail: publishing@cbspd.com; publicity@cbspd.com

Branches

- **Bengaluru:** Seema House 2975, 17th Cross, KR Road, Banasankari 2nd Stage, Bengaluru 560 070, Karnataka, India
  Ph: +91-80-26771678/79   Fax: +91-80-26771680   e-mail: bangalore@cbspd.com
- **Chennai:** 7, Subbaraya Street, Shenoy Nagar, Chennai 600 030, Tamil Nadu, India
  Ph: +91-44-26680620/26681266   Fax: +91-44-42032115   e-mail: chennai@cbspd.com
- **Kochi:** 42/1325, 26, Power House Road, Opp KSEB, Ernakulam 682 018, Kerala, India
  Ph: +91-484-4059061-65   Fax: +91-484-4059065   e-mail: kochi@cbspd.com
- **Kolkata:** 147, Hind Ceramics Compound, 1st Floor, Nilgunj road, Belghoria, Kolkata 700 056, West Bengal, India
  Ph: +91-33-25633055/56   e-mail: kolkata@cbspd.com
- **Lucknow:** Basement, Khushnuma Complex, 7-Meerabai Marg (behind Jawahar Bhawan), Lucknow 226 001, Uttar Pradesh, India
  Ph: +91-552-4000032   e-mail: tiwari.lucknowi@cbspd.com
- **Mumbai:** PWD shed, Gala No. 25/26, Ramchandra Bhatt Marg, Next to JJ Hospital Gate No. 2, Opp. Union Bank of India, Noorbaug, Mumbai 400 009, Maharashtra, India
  Ph: +91-22-66661880/89   e-mail: mumbai@cbspd.com

Representatives

- **Hyderabad** 0-9885175004
- **Jharkhand** 0-9811541605
- **Nagpur** 0-9421945513
- **Patna** 0-9334159340
- **Pune** 0-9923910676
- **Uttarakhand** 0-9716462459

Printed at: Mudrak, Nodia, UP India

Contributors

Core Workshop Faculty

Dr Arvind Bagga
Past Associate Editor, *Indian Pediatrics*
Professor, Department of Pediatrics
All India Institute of Medical Sciences
New Delhi 110029, India.
arvindbagga@hotmail.com

Dr Pooja Dewan
Associate Editor, *Indian Pediatrics*
Professor, Department of Pediatrics
University College of Medical Sciences
Delhi 110 095, India.
poojadewan@hotmail.com

Dr Piyush Gupta
Past Editor-in-Chief, *Indian Pediatrics*
Professor & Head, Department of
Pediatrics, University College of
Medical Sciences, Delhi 110 095, India.
prof.piyush.gupta@gmail.com

Dr Peeyush Jain
Executive Member, *Indian Pediatrics*
Specialist, Department of Pediatrics
Hindu Rao Hospital, North Delhi
Municipal Corporation, Malka Ganj
New Delhi, India.
peeyushjain@gmail.com

Dr Rakesh Lodha
Associate Editor, *Indian Pediatrics*
Professor, Department of Pediatrics
All India Institute of Medical Sciences
New Delhi 110029, India.
rlodha1661@gmail.com

Dr Devendra Mishra
Managing Editor, *Indian Pediatrics*
Professor, Department of Pediatrics
Maulana Azad Medical College
New Delhi 110 002, India.
drdmishra@gmail.com

Dr Anup Mohta
Associate Editor, *Indian Pediatrics*
Professor, Department of Surgery
Lady Hardinge Medical College
New Delhi, India.
mohtaanup@gmail.com

Dr Siddarth Ramji
Executive Editor, *Indian Pediatrics*
Director-Professor (Pediatrics)
Department of Neonatology
Maulana Azad Medical College
New Delhi 110 002, India.
siddarthramji@gmail.com

Dr Dheeraj Shah
Editor-in-Chief, *Indian Pediatrics*
Professor, Department of Pediatrics
University College of Medical Sciences
Delhi 110 095, India.
shahdheeraj@hotmail.com

Other Contributors

Dr Nitin Agarwal
Professor
Department of Surgery and Renal
Transplant
PGIMER, Dr. RML Hospital and GGSIPU
New Delhi, India.
dr.nitinagarwal76@gmail.com

Dr Shilpa Khanna Arora
Associate Professor
Department of Pediatrics, Post Graduate
Institute of Medical Education & Research
and Dr. Ram Manohar Lohia Hospital,
Baba Kharag Singh Marg, New Delhi
India. drshilpakhanna@yahoo.co.in

Dr. Shalu Jain
Senior Resident
Department of Pediatrics
University College of Medical Sciences
& GTB Hospital, Delhi 110 095, India.
shalu0927@gmail.com

Dr Jaya Shankar Kaushik
Editorial Board Member,
Indian Pediatrics
Associate Professor
Department of Pediatrics
Pt. B.D. Sharma PGIMS
Rohtak, Haryana, India.
jayashankarkaushik@gmail.com

Dr. Amir Maroof Khan
Adviser (Statistics)
Indian Pediatrics
Associate Professor
Department of Community Medicine
University College of Medical Sciences
University of Delhi, Delhi-95, India.
khanamirmaroof@yahoo.com

Dr Medha Mohta
Director-Professor
Department of Anesthesiology and
Critical Care
University College of Medical Sciences and
Guru Teg Bahadur Hospital Delhi, India.
medhamohta@gmail.com

Dr Sumaira Khalil
Specialist,
Department of Pediatrics
Safdarjung Hospital New Delhi, India.
sumairakhalil@yahoo.com

Dr Aparna Mukherjee
Department of Pediatrics
All India Institute of Medical Sciences
New Delhi, 110029, India.
aparna.sinha.deb@gmail.com

Dr AK Patwari
Former Associate Editor, *Indian Pediatrics*
Former Professor of Pediatrics, LHMC
New Delhi, and Research Professor
International Health, Boston University
School of Public Health, Boston
akpatwari@gmail.com

Dr HPS Sachdev
Past Editor-in-Chief, *Indian Pediatrics*
Senior Consultant, Department of
Pediatrics and Clinical Epidemiology
Sitaram Bhartia Institute of Science and
Research, New Delhi, India.
hpssachdev@gmail.com

Dr Deepika Upadhyay
Assistant Professor
Department of Commerce, Christ
University, Bengaluru, Karnataka, India.
deepika.upadhyay@christuniversity.in

Foreword

This is a dream come true! Words are failing me to express my happiness in writing the Foreword for the compilation on "The Art and Science of Writing a Scientific Paper." The concept was sown into my mind by my mentor and predecessor in Indian Pediatrics, Dr Panna Choudhury. And it took almost two years to bring it into a deliverable. *Indian Pediatrics* conducted the first workshop on "Art and Science of Paper-writing" in 2009 and since then there is no looking back. It stemmed from my firm belief that the editors' job is not only to accept/reject manuscripts; but training of researchers in the writing skills needed to get their papers published in reputed journals is also one of their mandatory duties.

Exposure to paper writing is almost non-existent in under-graduate medical curriculum in India; and limited to thesis writing in postgraduate courses. Little realizing that thesis and paper writing are different ball games altogether, most young researchers try to convert thesis into a paper by 'cut-paste' that usually results in rejection by the journals. Faculty selection and promotions are precariously hinged on the quantity and quality of their publications, but there are little avenues for structured training in paper-writing. *Indian Pediatrics* started this unique workshop and carried it to every corner of the country. As word of our workshop spread, we realized that we need to do beyond it, as the number of these workshop was probably too less to cover the target population. And thus the idea to translate the workshop sessions into written text and subsequently publish them as individual articles in *Indian Pediatrics,* emerged. This book compiles all these articles published in the Journal over a period of several months in single cover, thanks to the vision and effort of its editors.

This book not only provides useful tips to write the paper but also identifies the typical problems, including responding to reviewers, comments and how to deal with rejection. The content and writing style remain Simple, the key mantra of any educational endeavor.

I extend my best wishes to the editors of this book Prof. Devendra Mishra and Prof. Dheeraj Shah (also, the current Editor-in-Chief of

Indian Pediatrics), and all contributing authors for the success of this book. This is going to be a must-buy and must-read book for budding researchers, and serve the ultimate purpose of promoting publication of good research from the country.

Piyush Gupta MD, FAMS
Professor of Pediatrics and
Ex-Editor-in-Chief, *Indian Pediatrics* (2008–2013)
University College of Medical Sciences, Delhi

Preface

Indian Pediatrics has had the distinction of being one of the first journal to start the unique hands-on workshops on "The Art and Science of Writing a Scientific Paper" almost a decade ago, under the leadership of Prof. Piyush Gupta (Editor-in-Chief, *Indian Pediatrics*, 2008–2013). The core group of journal committee members of *Indian Pediatrics* have held many such workshops all over India since then, and benefited more than 600 participants. Further, for the benefit of millions of journal readers, the workshop facilitators were requested to write the core messages of their presentations in the form of articles for the journal. Given the continued popularity of these workshops, and the large number of web-hits on the articles, this book provides a compilation of these articles for the larger readership in an easily accessible form.

These book chapters reflect the experience gleaned from these flagship workshops of the journal, and would assist the novice medical researchers/authors in their publishing endeavors. We hope that even the experienced writer may be exposed to a few new tricks of the trade here. The annexures containing important information on critical aspects of the manuscript-preparation, would definitely add value to this book.

We hope this collection would also be of use as a reading material for workshops on scientific writing. Comments/suggestions are welcome at *jiap@nic.in (E-mail), www.facebook.com/indianpediatrics* (Facebook) or *@EditorIndPed* (Twitter).

We are also thankful to CBS Publishers & Distributors. We would like to put on record the sincere efforts of Mr YN Arjuna (Senior Vice President Publishing, Editorial and Publicity), and his team comprising Ms Ritu Chawla (GM Production), Mr Parmod Kumar, Mr Kshirod Kumar and Mr Rohan Prasad, for bringing out the book in the present form.

Devendra Mishra
Dheeraj Shah

Preface

Why Do We Write?

•HPS Sachdev •Siddarth Ramji

Have you ever considered why people write? Whether it is a professional need or a "hobby", it remains a powerful medium of expression of human thoughts and ideas. So is it with scientists and researchers; biomedical scientists being no exceptions. For academic health professionals, scholarship and publication are key activities.[1] Scholarship and publication can be expressed in a variety of ways, not limited to scientific journal articles alone.[1]

Writing and publishing scientific work can be a curiosity for some, a necessity for others, or an obsession or a passion for the rest. The reasons are dynamic depending on where one is on their career pathway.

PERCEPTIONS OF A DEBUTANT WRITER

A new entrant into the arena of biomedical research or practice is curious and wants to learn the 'art' of publishing scientific work. Novices need help and guidance to be initiated into scientific publication. Of course, it could carry the price of gift authorship to 'mentors', 'seniors' or 'guides'. But more importantly, seeing one's name in print boosts one's ego, improves one's acceptability amongst peers, and increases one's visibility in the scientific community. Ultimately, for those aspiring for an academic and research career, these are the first steps for career advancement in a ruthless and competitive academic world.

The forays into the publishing arena could be in the form of scientific abstracts in conference proceedings, case reports, scientific correspondence, interesting or novel images or chapters in books.

First published as Sachdev HPS, Ramji S. Why do we write? Indian Pediatr. 2016; 53:45–6.

MID-CAREER EXPECTATIONS

The reason why one publishes in their mid-career changes track slightly. Of course one continues to enjoy the 'name and fame' that publishing brings. It may be still important for progress up the academic ladder but the need to please one's 'bosses' may become less compelling. But there are more important reasons than just curiosity or necessity. There is deep desire to contribute to science, and share outcomes of original research with the scientific community. At this stage, one also may be keen to share updated information for appropriate practice to the medical community. One may also wish to play the mentorship role to budding young scientists. By this time one has honed the skills of good scientific writing, and one cannot but help being passionate about writing.

Scientific writing is not only a mode of communication, but also impacts other facets of human development. It helps to process one's thinking and become active learners.[2] For the physician, it also improves one's clinical skills—better clinical notations, structured approach to problem solving, better judgment—and teaches one to be more appreciative and accommodative to the suggestions of colleagues.

LATE-CAREER THOUGHTS

For most biomedical scientists, publishing would almost have become an 'addiction' at this stage of their career. One continues enjoying contributing to science and being a mentor to juniors. But above all, one of the greatest desires is to be able to influence practice and policy so that all, but more particularly the marginalized and under-privileged, are benefitted by the advances of medical science.

'UNDESIRABLE' REASONS

There are other reasons why one may write. While some of these reasons may be acceptable for popular science but the ethics of it for the advancement of biomedical research is debatable. Serving as "ghost writer" for money is unquestionably not ethical in scientific publication. Similarly, writing to promote a product or sell an intervention, or writing with the primary intent of promoting one's self or institution would also be considered undesirable. One should always be wary of falling into this trap with the increasing commercialization within the present day healthcare delivery system.

No doubt that those who publish flourish in the academic world. But not all are in that race, and for the large majority, publishing can be fun. But writing requires patience and perseverance. As James Hardee said *"It is a slow birthing process. But once you get your first one done, it is sort of addictive".*[3] No one is perfect; remember everyone has to sweat through the labor of publication. Many of us, who today may be considered successful because of the plethora of publications against our names, have had multiple rejections during our initial attempts at publication. But the reviewer's comments and criticism have actually improved the quality of one's scientific papers, and one remains ever grateful for their time and effort in improving the quality of the publications. As biomedical scientists, we must publish because the information belongs to all of mankind.

In summary, the reasons for writing a scientific publication can be several, and these may vary with the stage of your career development. If any of the above listed reasons entices someone who is a beginner, do not delay action but be bold enough to give it a shot! The subsequent publications in this series will be a helpful guide for embarking on this journey and also refining the skills.

REFERENCES

1. McGaghie WC, Webster A. Scholarship, publication, and career advancement in health professions education. Med Teach. 2009; 31:574-90.
2. A Message from the National Writing Projects of Michigan. Available from: *http://www.michigan.gov/documents/National_WP_151557_7.pdf.* Accessed December 6, 2015.
3. Christine Wiebe. Trying to Improve Your Clinical Presentation. ACP Observer, November 1997. Available from: *http://www.acpinternist.org/archives/1997/11/clinpress. htm.* Accessed December 5, 2015.

2

A Writer's Dilemma: Where to Publish and Where Not to?

•Pooja Dewan •Dheeraj Shah

Journals are the essence of scholarly communication. They not only serve to disseminate latest scientific advancements but also provide a platform for archiving scholarly information for future reference, and allow a researcher to assert his scientific mettle. Selecting the most suitable journal to showcase one's scholarly work is no mean feat. With more than 43,000 biomedical journals listed with PubMed,[1] the database maintained by United States National Library of Medicine (NLM), this exercise can easily flummox an inexperienced researcher. The huge risk of rejection of a paper from a journal that is not the right fit, and a widening web of dubious and predatory journals which publish almost everything sent to them, make this task particularly daunting.

WHERE TO PUBLISH?

The aim of clinical research is to bring about a positive change in practice and policy so that mankind is benefitted by the advances of medical science.[2] Therefore, unless the research work gets published and reaches its target audience, the entire exercise can be futile. Failure to choose the appropriate journal results in rejection and wastage of precious time, and slow career progress for the researcher. To facilitate the process of selecting the most appropriate journal, we need to consider the following variables (Box 2.1):

Focus: Every journal targets a certain audience and has a certain focus. On the basis of their focus, journals can be categorized as: broad-specialty *vs* specialty journals, pure research *vs* applied science journals, qualitative research *vs* quantitative research

First published as Dewan P, Shah D.A writer's dilemma: where to publish and where not to? Indian Pediatr. 2016; 53:141–5.

Box 2.1	**Variables to consider for choosing a journal**

- Focus/scope
- Indexing status
- Impact factor
- Peer-reviewed
- Affiliation to scientific societies
- Publication frequency
- Publication fees
- Accessibility
- Time to publication/early online version

journals, veterinary (animal) science *vs* human science journals, etc. Likewise, some journals may have a more local and regional appeal, while others may have a more global readership. Specialized journals, even with a potentially smaller readership, may disseminate your work more efficiently to your desired audience than a broad-specialty journal. It is important to remember that we should not only be interested in getting our work published, but also aim to get it noticed by the right audience. Therefore, it makes sense to publish data pertaining to a regional community in a local journal where it may influence the practice and policy rather than publishing in a 'reputed' international journal that is seldom referred to by the endasers. For example, a research work evaluating the predictors of mortality in children suffering from dengue fever in an urban belt in India would be better appreciated and read in a journal popular (and published) in India rather than a foreign journal with very limited circulation in the region of the origin of work.

The focus of the journal is usually stated on the journal's home page under the heading 'scope of the journal' or in the instructions to authors. A look at the recent issues of the journal will also give you an idea of the journal's area of focus. It is important to ascertain the harmony between the theme of the manuscript and focus of the journal before submission, as a mismatch between the two is one of the leading causes for outright rejection of the manuscript.

Indexing status: Indexing of a journal in a citation database is a property by the virtue of which articles published in it become searchable in that database.[3] The content published in the journal is indexed at the article level by assigning keywords, and then making them searchable in the database. Other bibliographic elements of journal articles, including authors' names, title of the article, journal name, and date of publication, are also used for indexing.

Index Medicus was the most widely accepted and comprehensive database of biomedical journals from 1879 until 2004. With the rapidly increasing number of journals, the printed publication Index Medicus was replaced by its online version Medline in 2005. Among the major databases for biomedical journals, indexing by Medline is considered as a benchmark of high quality for a journal. Over the years, other databases like Embase, Scopus, Science Citation Index, Directory of Open Access Journals, and many regional databases have emerged. Remember Google, Google Scholar and Sherpa-Romeo are not citation databases!

However, indexing of a journal comes with its own problems. Inclusion of a journal in a reputed indexing database depends on its scientific merit and rigorous publication policy and ethics, and therefore not all journals get indexed. Several regional and national journals, published in native languages, fail in their attempt to be indexed in the international databases. We must remember that not all research is relevant globally, and some may only be suited for publication in a regional or national journal that may not be indexed. Therefore, although important, indexing should not be used as the sole criterion for choosing the journal.

Despite the fallacies of indexing, it continues to be a major tool for assessing the merit of scientific publications. The recent Medical Council of India (MCI) guidelines recommend that publications indexed in Scopus, PubMed, Medline, Embase/Excerpta Medica, Index Medicus and Index Copernicus should be considered for promotions of teaching faculty in medical colleges.[4] This has generated considerable debate amongst medical fraternity as indexing databases like Science Citation Index and IndMed have been overlooked whereas a database with questionable integrity–Index Copernicus–has been included.[4] These MCI recommendations raise some important questions: 'Whether an indexed journal should be preferred over a non-indexed journal with a high potential of influencing change in practice and policy', 'Which indexing database is valid,' and 'Whether publications should be evaluated for scientific merit by indexing status of the journal rather than peer review?'

Impact factor: Another parameter, the impact factor (IF), is often used as a proxy for the relative importance of a journal within its field, and is frequently over-rated. IF of a journal is the annual measure of the extent to which articles published in that journal are cited. IF is awarded to the journals indexed in Science Citation Index, published annually in *Thomson Reuters Journal Citation Reports.*[5] However, IF must be interpreted with caution as its calculations are prone to manipulation.[6] Editorial policies such as preferential publication of review articles and articles dealing with newer diagnostics and therapeutics, short publication lag, and excessive self-citation can magnify the IF. English language journals and Basic sciences journals have higher impact factors. Abuse of the IF and the dominance of the prominent journals is a threat to the smaller and non-English language journals, and is akin to the 'Matthew effect' whereby the rich get richer and the poor become poorer. Interestingly, IF is not available for all indexed journals as not all journals indexed in MedLine are indexed in the Science Citation Index.[7] Moreover, the IF of a journal just tells about the merit of the journal, and not that of a particular article published in the journal.

Considering the potential problems in calculation of IF, it may be advisable to explore certain other bibliometric indices (Box 2.2) like Immediacy index, Cited half-life, SCImago journal rank and Eigenfactor score to compare journals.[8] Likewise, it is important to remember that for evaluating a researcher's academic merit, h-index, i-10 index, and citations (Box 2.3) are more relevant indices than the above.

Affiliation of the Journal to prestigious organizations: A journal publisher who is a member of the Committee on Publication Ethics (COPE) indicates that the Journal will follow the essential norms on publication ethics. COPE is a platform for editors and publishers of peer reviewed journals to discuss and seek advice on the ethical issues of publishing. Another indicator of the journal quality is its affiliation to the International Committee of Medical Journal Editors (ICMJE) which would also indicate that the journal abides by the publication recommendations given by them. Open access journals listed in the Directory of Open Access Journals (DOAJ) and Open Access Scholarly Publisher's Association also signify its credibility. Journals owned by reputed scientific societies (academies) are perceived to be superior.

Box 2.2 | **Bibliometric indices to assess the importance of a journal**

- *Impact factor.* The number of citations per paper received by a journal in a particular year to the papers published in that journal during the two preceding years.
- *Immediacy index.* The average number of times an article gets cited in the year it is published, and hence indicates how quickly articles in a journal are cited.
- *Cited half life of a journal.* The median age of the articles in the journal that were cited by other journals during the year. Therefore it reveals whether articles that were published a long time ago in that journal are still being cited.
- *Eigenfactor score.* This score evaluates journals according to the number of incoming citations over the preceding five years, with citations from highly ranked journals weighted to make a larger contribution to the Eigenfactor score than those from poorly ranked journals (Page rank algorithm).
- *SCImago Journal Rank (SJR).* It is a measure of scientific influence of scholarly journals that accounts for both the number of citations received by a journal and the importance or prestige of the journals where such citations come from. Calculation of the SJR indicator is very similar to the Eigenfactor score, with the former being based on the Scopus database and the latter on the Web of Science database
- *Altmetric.* It is a non-traditional broad group of metrics which cover other aspects of the impact of a work, such as how many data and knowledge bases refer to it, cite it, article views (PDF or HTML views), downloads, or mentions in media (journal comments, science blogs, Wikipedia, Twitter, Facebook and other social media).

Box 2.3 | **Bibliometric indices to assess the academic contribution of a researcher**

- *Citations.* Citations of a researcher are the total number of citations to all articles authored by the researcher.
- *Five-year citation.* Number of citations received by an author/article in last five years.
- *h-index.* The largest number h such that an author's 'h' publications have at least 'h' citations.
- *i10-index.* It is the number of publications with at least 10 citations.

Peer review: Peer review process is a service rendered by reviewers who provide honest and constructive criticism of research work to assess it worthiness for publication in a journal. Hence, peer review process is vital element of scholarly publishing and peer-reviewed journals are considered honorable.[9]

Reputation among colleagues: A simpler way to assess the reputation of a journal could be asking your peers or mentor about their choice of journal.

Accessibility: Journals which have both print and online versions have easier accessibility and hence may be preferred. Journals providing free online content are more accessible, especially to readers from underprivileged settings. In addition, regional or national journals with English-translated versions may be globally more acceptable.

Time-to-print: Many journals, including *Indian Pediatrics*, declare the date of initial submission and the date of final acceptance at the time of final publication. Journals offering a reasonable time frame for publication should be preferred, lest the research becomes outdated. However, with rampant unethical publishing practices, authors need to be cautious while choosing to publish in journals offering fast-track publication as many of these may actually be predatory (discussed later).

Format: To avoid outright rejection, one must check whether the journal has a policy of accepting articles of the form you are writing. This can be ascertained by reading the 'instructions to authors' of the journal, as well as looking at the past issues of the journal. It is important to ascertain whether the manuscript structure (order of sections), reference style, figure formats and image specifications match with the journal's style before submission. In case, you are not clear on this aspect, you can verify this with the editorial team by sending them an email (pre-submission enquiry). This may be particularly relevant in case of review articles as some journals may solicit them from experts in the field.

Publication charges: In the traditional publishing model, the access to published research work was controlled by the publishers who charged libraries, institutions and individuals a subscription fee and also a per article fee. This was referred to as the "green road" of publishing. This model led to frustration amongst individual researchers as they could not afford to pay the hefty journal subscriptions which witnessed a steady annual rise of 8–10%. However, 2002 witnessed the Open access (OA) movement in scholarly publishing wherein users could download and read journal articles on the internet without having to pay for it.[10] The publishers of the OA journals recovered costs by charging the authors a publication fee. This model is referred to as the "gold road" of publication. Since most health care research globally is public-funded, the OA model does seems righteous in allowing all researchers a free access to research and in assisting accelerated discovery and advancements in biomedical research. OA model also allows researchers to get more citations for their research. However, the economic viability of this model is debatable as the "author-pay" model is a major obstacle for researchers from developing countries who already struggle to get institutional budgetary allocation for research and hence opt out of publishing in OA journals unless granted a waiver of the author fees, which is usually tough. The publishers of OA journals also feel that author fees alone cannot sustain the high publication costs. While most OA journals do try to ensure a fair peer-review process to ensure high academic standards, lately there has been the emergence of several OA journals that compromise on peer review process and the quality of papers, with the aim of publishing more articles to generate more revenue from authors. Author fees, lack of journal prestige, ethical concerns, and loss of author copyright control are some of the major drawbacks of the OA model, and force many authors to tread the green road of publication.

WHERE NOT TO PUBLISH?

Beware: Predatory Journals on the Prowl!

For sustained academic growth, ethical publishing is a pre-requisite. With the plethora of biomedical journals to choose from, authors need to be discerning more than ever before. While the OA model in publishing fostered easy and free access to innovative high-quality scholarly research, there was also a flip side to it.[11] Several poor-quality journals emerged on the internet that exploited the OA model by offering fast-track publication of poor-quality research without any peer-review, in return for a nominal article processing fee.[12] While some of these journals state the publication fee upfront, most notify the publication fee to the author only after his/her manuscript is accepted for publication. By then, much time and effort has already gone into the review and revision of article, and the author has little option other than to pay the fee. These publishers lure naïve researchers—mostly from developing world—by sending them spam and phishing emails inviting them to publish research work in their journals. Their emails often display the phrase "CALL FOR PAPERS" in capital and large fonts. Some of them even strategize to entice authors by sending them personalized emails praising their recent scholarly work and inviting them to submit similar research work. In order to promote the credibility of their journals, these publishers request researchers to join their editorial teams as members and editors. Their mails often have several spelling and grammatical errors. This unabashed unethical seeking of authors by publishers was first pointed out by Jeffery Beall, an academic librarian from Colorado, USA, who christened them as 'Predatory publishers' in his blog in 2010. In 2011, Beall published a list containing the names of 18 predatory publishers,[13] now called as the Beall's list. In 2012, Beall shifted his blog to a Word Press platform and named it Scholarly Open Access (found at *http://scholarlyoa.com*). In his blog, he updated the list of 'Predatory Publishers' and "Predatory Journals" annually to caution inexperienced researchers and authors. Beall noticed that these publishers had certain common traits and postulated criteria to help identify them in his blog. Such publishers usually did not state the location of their headquarters and their websites were of poor quality, replete with typographical and grammatical errors. Papers in the publisher's journals were not only of inferior academic standards, but were also poorly copy-edited. These publishers had a large portfolio with several journal titles with most of them being recent with scant content. The publishers' emails have freely available domain names like gmail, hotmail, yahoomail, etc. Be all also noticed that these journals had little geographical diversity among editorial board members as well as authors, and the editorial board member list exhibited a male preponderance. The PDF files of papers published in these journals were locked to prevent them from being vetted for authenticity, and the publisher deliberately prevented the content from being indexed in academic indices. In addition, most of these journals adopted a nomenclature to closely mimic a reputed journal in the field, to beguile inexperienced authors. He cautioned against journals with ambitious titles containing terms like "Innovative", "World", "International", "Global", "European", and "Euro-Asian". Predatory Publishers have been known to make bogus claims about their indexing status and high impact factors of their journals.[14,15] They also seek the assistance of companies which claim to provide valid scholarly metrics such as CiteFactor, Advanced Science Index, General Impact

Factors, Global Impact Factors and Science Impact Factor. These bibliometric indices have been described by Beall as 'Misleading metrics' as their calculations are non-transparent, unscientific and manipulated. These companies charge the publisher and assign them indices which increase with time, in an attempt to mesmerize naïve researchers. Since 2011, there has been a deluge of predatory publishers, with 923 predatory publishers (publishing thousands of predatory journals) and 882 stand-alone predatory journals being listed in 2016.[16]

Another concept of 'Hijacked Journals' was also noticed by Beall, wherein a publisher creates a website which falsely claims to be the website of an authentic scholarly journal and even provides links to the bibliographic content of the authentic journal. They then invite manuscript submissions for the hijacked version of the journal and make away with the article submission and processing fees. In 2016, Beall's list mentioned 101 hijacked journals against 30 such journals listed in 2015.

We advise authors to ascertain the credibility of journals by checking their credentials, indexing status, open access and archiving options, affiliation to scientific societies, the reputation of the publisher, and the Beall's list. The last is particularly important, as any association with predatory journals is now being viewed negatively by the scientific world.

With the groundwork for publication done, in the next chapter we move onto the 'How' of writing a paper for a scientific journal.

REFERENCES

1. US. National Library of Medicine. List of all journal cited in Pubmed®. Available from: *https://www.nlm.nih.gov/bsd/serfile_addedinfo.html*. Accessed January 9, 2016.
2. Sachdev HPS, Ramji S. Why do we write? Indian Pediatr. 2016; 53:45–6.
3. Ng KH, Peh WC. Getting to know journal bibliographic databases. Singapore Med J. 2010; 51:757–60; quiz 761.
4. Aggarwal R, Gogtay N, Kumar R, Sahni P, for the Indian Association of Medical Journal Editors. The revised guidelines of the Medical Council of India for academic promotions: Need for a rethink. Indian Pediatr. 2016; 53:23–6.
5. Garfield E. The history and meaning of the journal impact factor. JAMA. 2006; 295:90–3.
6. Lippi G, Favalor EJ, Simundic AM. Biomedical research platforms and their influence on article submissions and journal rankings: an update. Biochem Med (Zagreb). 2012; 22:7–14.

7. Elsaie ML, Kammer J. Impactitis: The impact factor myth syndrome. Indian J Dermatol.2009; 54:83–5.
8. Brown T. Journal quality metrics: Options to consider other than impact factors. Am J Occup Ther. 2011; 65:346–50.
9. Hojat M, Gonnella JS, Caelleigh AS. Impartial judgment by the "gatekeepers" of science: fallibility and accountability in the peer review process. Adv Health Sci Educ Theory Pract. 2003; 8:75–96.
10. Albert KM. Open access: implications for scholarly publishing and medical libraries. J Med Libr Assoc. 2006; 94:253–62.
11. Bowman JD. Predatory publishing, questionable peer review, and fraudulent conferences. Am J Pharm Educ. 2014; 78:176.
12. Beall J. Predatory publishers are corrupting open access. Nature. 2012; 489:179.
13. Beall J. Medical publishing triage-chronicling predatory open access pmublishers. Ann Med Surg (Lond). 2013; 2:47–9.
14. Gasparyan AY, Yessirkepov M, Diyanova SN, Kitas GD. Publishing ethics and predatory practices: A dilemma for all stakeholders of science communication. J Korean Med Sci. 2015; 30:1010-6.
15. Butler D. Investigating journals: The dark side of publishing. Nature. 2013; 495:433–5.
16. Beall J. Beall's List of Predatory Publishers 2016. Available from: *http://scholarlyoa.com/2016/01/05/bealls-list-of-predatory-publishers-2016/*. Accessed January 9, 2016.

SNIPPET FROM THE ARCHIVES

Time Lag from Submission to Printing

(Extracted from Bagla J, Mishra D. Time-lag from submission to printing in Indian biomedical journals. Indian Pediatr. 2011; 48: 67–8.)

Research manuscripts face a 12–18 months time-lag from initial submission to final publication in a scientific journal. The time lag in publication may adversely impact the careers of younger scientists, in addition to loss of information by hindering timely incorporation of major advances into the policy and practice of medicine.

The timeliness of publication in five Indian, clinical, biomedical journals (Indian Pediatrics, the Indian Journal of Pediatrics, Neurology India, The Indian Journal of Medical Research and Journal of Postgraduate Medicine) was compared for a two-year period. The time from manuscript submission to publication for the journals studied (median: 358.3 days; range: 202.9-421.3 days) was not significantly different, but was much higher than that for many international journals.

Table: Time taken for publication of original papers (2007–2008)

Mean time (d)	Indian Pediatr (n=127)	Indian J Pediatr (n=146)	Indian J Med Res (n=188)	*J Postgrad Med (n=40)	*Neurology India (n=60)	All journals
Submission to review	83.5	–	–	107.6	–	91.6
Review to acceptance	169.9	–	–	36.7	–	125.5
Submission to acceptance	253.6	211.1	–	144.3	–	215.6
Acceptance to publication	130.2	146.4	–	58.6	73.2	107.7
Submission to publication	383.7	358.3	421.3	202.9	–	350.0

Indian Pediatr: Indian Pediatrics; Indian J Pediatr: Indian Journal of Pediatrics; Indian J Med Res: Indian Journal of Medical Research; J Postgrad Med: Journal of Postgraduate Medicine; All were monthly publications except *quarterly publications.

3

Writing the Title, Abstract and Introduction: Looks Matter!

•*Pooja Dewan* •*Piyush Gupta*

"I always avoided eating the pineapple due to its tough and spiky exterior, until a friend of mine offered me a slice of it." That friend may not be there at all times for all of us; therefore, you need to ensure that what catches the eye first, needs to be inviting. The Title, Abstract and the Introduction are the face of any research paper, and hence need to be dressed in such a way as to enthrall the readers.[1]

GETTING THE TITLE RIGHT

The title is the first part of any manuscript that is seen by the editors, reviewers, as well as readers. It is also what appears on the contents page of the journal issue, and serves as a window to the research paper.[2] A strong title pulls readers into it, making it memorable and encouraging for people to read. A weak title dulls the readers' expectations and could negatively affect the views on your research work, no matter how good it is.

Most electronic databases and search engines, and journal websites, use the words in the title or the keywords provided by the authors to retrieve the scientific paper during online searches. Therefore, title plays a crucial role in ensuring the access of your paper to its readers. Busy editors often decide the eligibility of a manuscript for publication and peer-review based on their initial impression after a scrutiny of the manuscript's title and abstract.[3] Therefore, having a good and inviting title should be a priority.

HOW TO WRITE A GOOD TITLE?

Keep it Concise

If the title is too long or complicated, it may put off the readers right at the onset. Use of about 10–12 words in the title will enable

First published as Dewan P, Gupta P. Writing the title, abstract and introduction: Looks matter! Indian Pediatr. 2016; 53:235–41.

you to bring out the essence of the research work (patient/species, intervention done, any comparisons, and outcome). Consider the following title:

"A novel study on the usefulness of NS1 antigen detection test in the diagnosis of dengue fever in children: analysis of clinical features and comparison with ELISA test and viral culture with clinical follow-up in 100 patients of dengue fever at XYZ Hospital, Delhi." This would take ages to read. Not many people will have the patience to go through this with a clear head! Now consider: *"NS1 antigen test for the diagnosis of Dengue fever"*. Obviously, this title is better because it is clear and concise. It permits the reader to proceed onto the next section within his/her attention span. To make the title concise, we need to avoid unnecessary phrases. Consider the following titles:

- <u>Role</u> of steroids in aplastic anemia
- <u>Effects</u> of antenatal exercises on birthweight of baby
- A <u>study</u> on efficacy of beta-blockers in heart failure

The underlined words in these titles do not add to the information provided, and by simply omitting these superfluous words, the title is as informative, and definitely sharper.

Keep it Specific

Let us consider another title: *"Vitamin D and Pneumonia."*

Despite being extremely concise, this title is still lacking the power to engage the reader as it is too general and vague. It does not lead the reader in any particular direction. Instead, it leaves the informative work to the abstract and the paper itself, which, as we know, not many people go over. Consider replacing it with *"Vitamin D deficiency and risk for severe pneumonia in under-five children."* This is longer but definitely more specific.

Whether to include Place of Study

Sometimes, a given study, if conducted with the same methodology, by the same researcher but in a new setting, may yield completely different results. Consider a study on prevalence of hypertension in young adults in Mumbai. Here the location of the study is vital to the study itself. The prevalence of hypertension at a certain geographical location is dependent on its prevalent lifestyle habits, which in turn are affected by the economic status and cultural and social practices. So, inclusion of the place of study in the title for this study would be desirable for sake of completing of information.

Now consider the following titles:

- *"Daily vs. weekly iron supplementation in adolescent girls in Delhi"*
- *"Methylprednisolone vs. Cyclosporine for treating Childhood Aplastic Anemia in Manchester"*

The study of iron supplementation in adolescent girls in Delhi will not be very different from the study of iron supplementation in adolescent girls elsewhere. The affecting factor here is not the socio-economic or political environment, but evolutionary build-up of a species, which will not differ even if we change the geographical location of the study. Same applies to the second study. The results obtained in the Manchester study are also applicable to other geographical locations. In such studies, the name of place becomes redundant in the title.

Placing the Keywords towards the Beginning

The important words and terms related to your study should be placed towards the beginning of the title. For example, *"Rituximab for Treatment of Autoimmune Hemolytic Anemia"* is a better title than *"Treatment of Autoimmune Hemolytic Anemia with Rituximab"*.

Let us take the example of a study being conducted to ascertain the differences in the prevalent trends of obesity between men and women. The title for this study can be composed in two ways: *"Prevalence of Obesity in Adults by Gender"* or *"Gender Differences in Prevalence of Obesity in Adults"*. Both titles are concise, specific, and bereft of unnecessary phrases, yet these are inherently different in their approach. In this example, the focus of the study is not prevalence of obesity *per se*, but the male-female comparison of prevalence of obesity. Therefore, the second title, which emphasizes the focus of the study by placing it in the beginning, is more appropriate.

Use of Colon in the Title

It is important to note that the study design is usually preceded by a colon in the title. For example, *"Azithromycin for treatment of enteric fever: a randomized controlled trial"*.

Use a Descriptive/Neutral Title

A descriptive title has all the elements of the research work (patient, intervention, outcome, comparison), yet it does not reveal the main findings of the study or its conclusion. Using too amusing or loud titles should be avoided and as far as possible use a neutral title.[4] For example, *"Seven Days versus Ten Days Antibiotic Therapy for Culture-Proven Neonatal Sepsis: A Randomized Controlled Trial"*.

Avoid Declarative Titles

A study title which states the main findings of the study is said to be a declarative one. It reflects the intrinsic bias on the part of the researcher regarding the interpretation of the data.

"Seven Days Antibiotic Therapy is better than Ten Days Antibiotic Therapy for Culture-Proven Neonatal Sepsis: A Randomized Controlled Trial" is a declarative way of writing the title for the previously mentioned study.

"Cryptosporidium is the Most Common Enteric Pathogen in HIV-infected Children with Diarrhea" is another example. *"Prevalence of Cryptosporidium in HIV-infected Children with Diarrhea"* is a more appropriate title as it lets the reader approach the subject with an open mind and retains the curiosity of the reader.

Avoid Query/Interrogative Titles

Introducing the subject of research in the form of a query can be distracting, and is best avoided. Consider the query version of the previous example: *"Is Seven Days Antibiotic Therapy Better Than Ten Days Antibiotic Therapy for Treating Culture-proven Neonatal Sepsis?"*

Query titles tend to sensationalize the subject and can sometimes be used for review articles. A research by Jamali and Nikzad[5] revealed that articles with query titles tend to get downloaded more frequently, yet they are cited less frequently.

Avoid Abbreviations/Acronyms in the Title

As far as possible, refrain from using abbreviations/acronyms in titles. Consider the title: *"Diagnosis of ARF in Children"*. Here, the abbreviation ARF could imply acute renal failure or acute rheumatic fever, and hence abbreviations are best avoided in titles. Now consider another title: *"IVIG for treatment of PANDAS"*. Here, IVIG is used for intravenous immunoglobulin and PANDAS denotes

Pediatric Autoimmune Neuropsychiatric Disorders associated with Streptococcal Infections. A reader unaware of their meaning may skip this article altogether.

However, abbreviations are sometimes useful for long, technical terms in scientific writing. The use of abbreviations that appear as word entries in Webster's Collegiate Dictionary is acceptable. For example: Use of abbreviations like HIV, AIDS, NADPH, and ATP, may be acceptable.

Ingredients of a Good Title

A balanced title needs to be "SPICED".[6] The acronym here refers to the six key elements of a title, i.e. Setting, Population, Intervention, Condition, End-point, and Design.

Setting: This refers to the situation in which the research takes place in. It could be community-based, home-based, school-based, hospital-based, or laboratory-based. Within the hospital itself, it could be amongst out-patients or inpatients, or in the emergency room. Likewise, it could be a rural or urban setting. It is important to mention the setting in the Title if results are not generalizable to other settings, or if the setting reflects the magnitude of your research. For example: *"Mortality in Severe Acute Malnutrition in Under-five Children: A Hospital-based Study."* Here it is important to mention the setting because mortality in severe acute malnutrition will be different in under-five children admitted to the hospital and those in the community.

Population: The population is the target of the research work and needs to be explicitly stated (age and/or sex, where necessary). For example: *"Prevalence of Depression in the Elderly"* and *"Prevalence of Osteoporosis in Post-menopausal Women."* In the first title only age is specified because sex may not be important. The latter title includes both age and sex, because of their relevance.

Intervention: Intervention (therapeutic or preventive) is a key element of any clinical trial. For example *"Vitamin D Supplementation in Children with Severe Asthma: A Randomized Controlled Trial"*. The study here could evaluate the effect of supplementation of Vitamin D on the severity of the asthma episode (therapeutic effect) or the occurrence of recurrent episodes of asthma (Preventive). The title should be able to clarify the type of study (*see* Design below) and

---

Box 3.1 | **Seven secrets to writing the title of a research paper**

- Keep it Concise.
- Keep it Specific.
- Decide regarding Place of Study.
- Use a Descriptive/Neutral title.
- Appropriately 'SPICED content.
- Avoid Interrogative or declarative titles.
- Avoid acronyms/abbreviations in the title.

---

the type of intervention, if it was planned. A still better title would be *"Therapeutic Effect of Vitamin D Supplementation in Children with Severe Asthma: A Randomized Controlled Trial"*.

Sometimes, research may only be observational with no intervention whatsoever. For example, *"Serum Vitamin B_{12} Levels in Adolescent Indian Girls: An Observational Study"*.

Condition: It refers to the clinical condition of the subjects. *"Serum Folate Levels in Pregnant Indian Women: An Observational Study"*, here the condition is pregnancy.

Endpoint: Outcome is sparingly used in the title, unless we wish to use a declarative title. It refers to the change or type of change the condition undergoes after being subjected to intervention.

Design: Including the study design in the title itself makes the title complete and it is usually placed after a colon or an em dash.

Box 3.1 summarises the tips for writing a good title.

When to Write the Title?

A rough title should be written when planning a study, keeping in mind at least five of the SPICED elements (except endpoint). This can be tailored after the completion of the study depending upon what we wish to highlight from the study.

CRAFTING A RUNNING TITLE

Many journals ask for a "running title" or "running head" or "short title" to be included in the submitted manuscript. This an abridged form of the main title, which is usually placed at the top left in the header of the published page of an article. The running title enables the reader to keep track of the article as he goes through loose printed pages of the article. Most journals would ask for a running title of no more than 50 characters including the spaces. To make the title still shorter, standard abbreviations could be used, and articles and study design be omitted. For example, the running title for a research paper titled *"Pulse Oximetry Screening to Detect Cyanotic Congenital Heart Disease in Sick Neonates in a Neonatal Intensive Care Unit"* can be written as *"Pulse Oximetry Screening for Cyanotic CHD"*.

CHOOSING THE KEY WORDS

The keywords you choose are important as these are used for indexing purposes.[7] Keywords are listed below the abstract text. It is important to not duplicate the "keywords" and "words used in the main title" as both enable accession and hence citation of your research work. Using the right keywords will speed up the internet retrieval of your work.[8] In order to determine the keywords, read through your paper and list the terms, phrases and abbreviations used frequently. Try to include variants of a term/phrase already used in your title as keywords, e.g. sepsis and septicemia, renal and kidney, tumor and cancer. Now refer to an indexing standard like the Medical Subject Headings (MeSH) database of the US National Library of Medicine.[9] Check if these terms are listed therein. The MeSH uses two tools to determine keywords:

- MeSH on demand
- MeSH browser.

'MeSH on demand' is a simple tool available from *https://www.nlm.nih.gov/mesh/MeSHonDemand.html*, which automatically deciphers the keywords from text such as an abstract or summary. MeSH Browser is a tool available from *https://www.nlm.nih.gov/mesh/MBrowser.html*, which allows for searches of MeSH terms, text-word searches of the Annotation and Scope Note, and searches of various fields for chemicals. Another way to identify keywords is to search similar research work from PubMed and then ascertain the MESH headings assigned to them. The keywords are not necessarily single words but may be two words. For example, "breast cancer" is listed as keywords in MeSH.

Before you finally submit your article, check if the keywords are appropriate. Type the keywords into the search engine and see if the search results resemble your research work.

WRITING THE ABSTRACT

The abstract is a concise statement of the major elements of your paper. It is usually the last section written by the authors, but is the first section of your paper that is read by the editors and reviewers. It should therefore provide a snapshot of the research undertaken by you. In addition, it should be comprehensive yet crisp.[10] The abstract should highlight the selling point of your research work and should lure the readers to read the whole paper. Besides determining selection of the paper, abstracts are also important for indexing. When searching online for research work, most databases would display the title as well as the abstract. This would enable the readers to determine if they really need to go through the full text of your research paper. However, it is important to remember that for your article to be picked up during an online search, it must contain the key words that a potential researcher would use to search.

What Should the Abstract Contain?

The abstract should be a window to your research and should effectively convey all the elements of the research work. The abstract essentially has four elements (Box 3.2).

Abstracts cover all the aspects of the research including the background, objectives, methods, results, conclusions, and recommendations. They, however, do not provide a critique of the research. These are usually around 300 words or about 10% of the length of the manuscript.

Format of an Abstract

Abstract can be written in running text without the use of subheadings (unstructured abstract) or it may be in a structured format with use of subheadings. A structured abstract may be a 4-point abstract or a detailed traditional 8-point abstract. A 4-point abstract has four subheadings, usually, (1) Background and/or

Box 3.2	Elements of the abstract
Purpose	Why this work? What was aimed?
Methods	How was it achieved?
Results	What are the findings?
Conclusion	What is the inference?

Objectives, (2) Methods, (3) Results, and (4) Conclusions. An 8-point abstract has eight subheadings, *viz*, (1) Objectives, (2) Study-design, (3) Study-setting, (4) Participants, (5) Methods/Intervention, (6) Outcome measures, (7) Results, and (8) Conclusions. You will have to choose the format of the abstract after checking the "Instructions to Authors" of the journal you wish to submit your research to.

The 4-point abstract is easy to write as the elements are distinct entities. They are:

Background: It should be brief and limited to two or three sentences, where you need to specify what is already known and why you conducted the study. The objectives of study should also be mentioned.

Methods: This is usually the longest section of the abstract and should give enough information to the reader to understand what and how was your study done. The important aspects that need to be covered here include the study design, study setting, clinical diagnosis of participants, sample size calculation, sampling methods, intervention done, duration of the study, research instruments used, and define the primary and secondary outcome measures and how these were assessed.

Results: This is the most important and difficult section to write in an abstract. The results should mention the exact number of participants including the drop outs, and adverse effects, if any. The results of the analysis of the primary objectives and the salient secondary objectives should be presented in words as well as numbers including P values. An abstract should present the results of your research as data (mean, standard deviation, 95% confidence interval, mean difference, P value, median, and interquartile range, where applicable). Merely stating the interpretation of results in sentences without numerical results is inappropriate. Consider the following examples:

"*Response rates differed significantly between hypertensive and non-hypertensive children.*" A better way to state your results is "*The response rate was higher in non-hypertensive than in hypertensive children (50% vs 20%, respectively; P <0.01).*"

Another example where results of a study evaluating the role of probiotics in diarrhea are presented:

"The time for resolution of diarrhea, and the recovery in terms of resolution of diarrhea and need for hospitalization was similar in the probiotic and placebo groups."

Or

"The median time for resolution of diarrhea was 54 hours in both the probiotic and the placebo group. Recovery in the probiotic group was marginally better but not statistically significant for resolution (hazard ratio = 0.91, 95% CI 0.60–1.31), rehydration (hazard ratio = 0.91, 95% CI 0.64–1.39) and hospitalization (hazard ratio = 0.94, 95% CI 0.67–1.34)."

The latter way of presenting the results in the abstract is better and more informative.

Conclusions: The conclusions need to state the 'take home message' and any other salient findings which need to be considered. The conclusions must always take into account your hypothesis and research question and must be written so as to answer the same in the light of your results. Additionally, you may present your perspective in this section of the abstract. Box 3.3 depicts some examples of writing the conclusions section of the abstract.

The 8-point abstract ensures completion of all aspects of the research; however, there is a significant overlap between the methods and results section. Therefore, it needs to be drafted very carefully. For example, under the subheading "participants", you will not only need to specify the inclusion/exclusion criteria (part of methods section), but also need to mention the exact number recruited in your study (part of results).

Box 3.3 | **Examples of the conclusions section of abstract**

- Daily zinc supplementation (40 mg for 14 days) in children aged five to 12-year with acute dehydrating diarrhea did not shorten the duration of diarrhea or reduce subsequent episodes. Further community-based trials with adequate sample sizes are needed.
- Pyomyositis is a specific and potentially fatal infection, which is common in India and must be differentiated from intermuscular abscess. A high index of suspicion and early institution of specific antibiotics followed by surgery can be lifesaving.

Attributes of a Good Abstract

A well-written abstract is characterized by the four Cs, *viz.* it should be complete, concise, clear and cohesive.

A good abstract should be *complete*. It should be a stand-alone document and cover all the major parts of the research in addition to bringing out its novelty.

A good abstract should be crisp and free from excessive wordiness or unnecessary information. For example, "X stimulates Y" will be a better choice of words than "X produces a stimulatory effect on Y". A good abstract should avoid too much background information. You should refrain from using empty phrases like "It was interesting to note that …". Cliché statements like "More research is needed" should be avoided. If there are implications, then you must state them clearly.

Box 3.4 **Example of a cohesive abstract[11]**

Vitamin D supplementation for severe pneumonia–a randomized controlled trial.

Objective: To determine the role of oral vitamin D supplementation for resolution of severe pneumonia in under-five children.

Design: Randomized, double blind, placebo-controlled trial.

Setting: Inpatients from a tertiary care hospital.

Participants: Two hundred children [mean (SD) age: 13.9 (11.7) months; boys: 120] between 2 months to 5 years with severe pneumonia. Pneumonia was diagnosed in the presence of fever, cough, tachypnea (as per WHO cut-offs) and crepitations. Children with pneumonia and chest indrawing or at least one of the danger sign (inability to feed, lethargy, cyanosis) were diagnosed as having severe pneumonia. The two groups were comparable for baseline characteristics including age, anthropometry, socio-demographic profile, and clinical and laboratory parameters.

Intervention: Oral vitamin D (1000 IU for <1 year and 2000 IU for >1 year) (n =100) or placebo (lactose) (n=100) once a day for 5 days, from enrolment. Both the groups received antibiotics as per the Indian Academy of Pediatrics guidelines, and supportive care (oxygen, intravenous fluids and monitoring).

Outcome variables: Primary—time to resolution of severe pneumonia. Secondary: duration of hospitalization and time to resolution of tachypnea, chest retractions and inability to feed.

Results: Median duration (SE, 95% CI) of resolution of severe pneumonia was similar in the two groups [vitamin D: 72 (3.7, 64.7–79.3) hours; placebo: 64 (4.5, 55.2–72.8) hours]. Duration of hospitalization and time to resolution of tachypnea, chest retractions, and inability to feed were also comparable between the two groups.

Conclusion: Short-term supplementation with oral vitamin D (1000–2000 IU per day for 5 days) has no beneficial effect on resolution of severe pneumonia in under-five children. Further studies need to be conducted with higher dose of Vitamin D or longer duration of supplementation to corroborate these findings.

The abstract should be *clear*, i.e. readable, well-organized, and not too jargon-laden. The abstract should be written in the past tense. Abstract written in active voice provides greater clarity. So, we may write "We conclude that ..." instead of "It was concluded that...". The findings of your research should not be discussed in the abstract, and any discussion should only be done in the main text of your research paper. The abstract should be free of figures, diagrams, tables, or images. The abstract should not contain any references/citations. Avoid use of abbreviations or acronyms.

A well written abstract should be *cohesive* and the text should flow smoothly between the parts. The abstract must follow the chronological order of sections in your main research paper ensuring a smooth transition. It must read like a story. A direct cohesiveness needs to be maintained between objectives, main outcome measures, results and conclusion (example in Box 3.4).

Before you finally submit your abstract check it for *consistency*; a mismatch between the abstract and main text may raise doubts on authenticity of your results. Check if the abstract meets the guidelines to authors in terms of format, word count, etc.

WRITING THE INTRODUCTION

Need for an Introduction

The introduction should aim to set the mood for your research, acquaint the readers with your research hypothesis, and motivate the readers to read your paper.[12] It should steer the readers from why you are doing the research into how you are going to fill the knowledge gap, i.e. into the methods section.

An introduction essentially has three main elements:

1. What is known?
The background of the research topic needs to be stated right at the onset to enable the readers to understand what is already known on the subject. This sets the stage for the basis of your research.

2. What is lacking?
You need to justify "why you are carrying out that research work", i.e. whether you are building upon previous research, looking at a novel aspect not evaluated by previous research, or if you are trying to improve upon a previous research that yielded ambiguous results.

3. What you aim to do?

You need to briefly state the objectives of your research. It is also advisable to present a detailed hypothesis at this juncture only.

How Long is too Long?

There are no strict word limits for writing the introduction; generally it is one of the shorter sections of the paper. Having the readers meander through too much of introduction can be counterproductive as it may cause them to lose focus and interest. You should assume that your work is going to be read by someone who has at least a reasonable knowledge about your research topic, so it is preferable that you do not beat about the bush. For example, for a study evaluating the role of probiotics in acute diarrhea in children, there is no need to discuss definitions and etiology of diarrhea in the introduction; you could start by commenting upon the well-established treatment options for acute diarrhea and how your study will add to the existing knowledge and practice.

How to Write the Introduction?

It would be useful to structure your introduction like an *"inverted pyramid"* or what could be simply said as *"funnel approach"*. This implies introducing the topic of the paper and discussing it in a broad context and then finally narrowing down to the research problem and hypothesis.

The introduction can be written in about two-to-three paragraphs. The opening paragraph should be dedicated to introducing the topic of research; it may also provide an overview of the topic of research. You must remember that the introduction is not a review of literature but it should convince the readers that you have thoroughly researched the topic and built their confidence in your hypothesis. A thorough literature search is an essential pre-requisite for identifying and framing the research question. However, a very lengthy literature review can put-off the readers so it is important to summarize what research has already been done on that topic, and highlight the lacunae or controversies regarding the same.

In the second paragraph, you need to identify a research niche. This can be done by highlighting the lacunae in existing research or opposing an existing practice or assumption. This will help you to arrive at your research question. You need to emphasize what additional knowledge will be gained through your research and how you aim to bridge the gap in knowledge. An ideal study should focus on a central question and may be another two or three questions that can be additionally addressed through your study. It is preferred to use "open-ended research question". A good research question should yield a testable hypothesis. It may be necessary for you to clarify any key terms or concepts in the introduction itself, particularly if you are dealing with an unfamiliar or new concept. It is also pertinent to declare any assumptions you are going to make in the research work.

In the third paragraph, you need to articulate your objectives and hypothesis. The hypothesis should be a tentative prediction of relationship between two or more variables. It should be neither too general nor too specific, and is often declarative. While stating the hypothesis it would be better to state it implicitly rather than saying that "Our research is based on the hypothesis ...". For example, a research hypothesis can be stated as "10-days duration of intravenous antibiotics is not inferior to 14-days therapy for treating neonatal septicemia". The hypothesis should be used to convince the readers about what results are expected from your research. Also, remember that a hypothesis is valuable even if proved to be wrong.

And as the phrase goes "Well begun is half done", so is the story with a research paper. A well drafted abstract and introduction section with a strong title will help the researcher to win half the battle.

REFERENCES

1. Sharp D. Kipling's guide to writing a scientific paper. Croat Med J. 2002; 43:262–7.
2. Peh WC, Ng KH. Title and title page. Singapore Med J. 2008; 49:607–8; quiz 609.
3. Fox CW, Burns CS. The relationship between manuscript title structure and success: editorial decisions and citation performance for an ecological journal. Ecol Evol. 2015; 5:1970–80.
4. Sagi I, Yechiam E. Amusing titles in scientific journals and article citation. J Information Science. 2008; 34:680–7.
5. Jamali HR, Nikzad M. Article title type and its relation with the number of downloads and citations. Scientometrics. 2011; 88:653–61.
6. Gupta P. Framing a suitable title. In: Gupta P, Singh N (Eds). How to Write the Thesis and Thesis Protocol. A Primer for Medical, Dental and

Nursing Courses. First Edition. New Delhi: Jaypee Brothers Medical Publishers; 2014. p. 45–9.

7. Darmoni SJ, Soualmia LF, Letord C, Jaulent MC, Griffon N, Thirion B, *et al*. Improving information retrieval using Medical Subject Headings Concepts: a test case on rare and chronic diseases. J Med Libr Assoc. 2012; 100:176–83.

8. Kabirzadeh A, Siamian H, Abadi EB, Saravi BM. Survey of keyword adjustment of published articles medical subject headings in journal of Mazandaran University of Medical Sciences (2009-2010). Acta Inform Med. 2013; 21:98–102.

9. Neveol A, Shooshan SE, Humphrey SM, Mork JG, Aronson AR. A recent advance in the automatic indexing of the biomedical literature. J Biomed Inform. 2009; 42(5):814–23.

10. Alexandrov AV, Hennerici MG. Writing good abstracts. Cerebrovasc Dis. 2007; 23:256–9.

11. Choudhary N, Gupta P. Vitamin D supplementation for severe pneumonia—a randomized controlled trial. Indian Pediatr. 2012; 49: 449–54.

12. Peh WC, Ng KH. Writing the introduction. Singapore Med J. 2008; 49: 756–7; quiz 758.

PEARLS OF WISDOM

Preparing the Title Page

Most journals require the authors to provide a Title page, which needs to mandatorily have information related to important aspects of the manuscript–even though, all this information may not go into the final article that is published (Box). We list herein some of the important considerations while preparing the title page:

Author names: Each author's name should be provided in the form of a first name and surname (both are essential). Due to the requirements of the indexing agencies, it is usually not possible to accept initials as surnames, e.g. if Vidya K is provided as the author name; 'K' will be considered as the initial and 'Vidya' will be indexed as the last name. Thus, authors wanting a specific part of their name to be initialized after publication should mention the same in the covering letter or along with the full name on the title page.

Word count: Some journals ask for the same—take care to note that it does not include abstract, tables, figures, acknowledgments, key messages and references.

Authors' contributions: A succinct description of what each author contributed should be provided on the title page. Statements like "all authors were involved in all aspects of manuscript preparation and submission" are usually not accepted. The name of the author who should be approached for access to raw data should also be stated (if different from the corresponding author). Most journals do not entertain requests for joint corresponding authorship or joint first authorship.

Acknowledgments: Any person who contributed to the work but does not satisfy all the conditions for authorship should be named in the acknowledgments e.g., providing purely technical help, writing assistance, or a department head who provided only general support. Groups of persons who have contributed materially to the paper but whose contributions do not justify authorship may be listed under a heading such as "clinical investigators" or "participating investigators," and their function or contribution should be described e.g., "served as scientific advisors," "critically reviewed the study proposal," "collected data," or "provided and cared for study patients." Authors should obtain written permissions from everyone acknowledged by name. Statements like "we thank all patients and their families" or "we acknowledge the help of all research staff" or "we thank the reviewers" are usually not preferred.

Funding: Authors should report all financial and material support for the research work, including grant number and funding agency.

Essential components of the title page

- Category of the manuscript
- Title of the paper (and a short running title) (see Chapter 3)
- Name of authors with highest academic degree
- Designation and affiliation of all authors (at time of conduct of the study)
- Corresponding author name and contact details (address, e-mail and phone)
- Acknowledgements
- Contribution of each author (see Chapter 8)
- Conflict of interest statement (see Chapter 8)
- Funding information

Additional details like word count, number of figures and tables, trial registration number, ethical clearance information, etc. may be required by different journals, and instructions for authors of the concerned journal should be consulted.

Devendra Mishra

Writing Methods: How to Write What You Did?

• Shilpa Khanna Arora • Dheeraj Shah

The section on methods is the most vital part of a study as well as of a manuscript. It is the most critically evaluated section of a paper not only by the reviewers but often by the readers as well. It should always be written in a simple and clearly understandable language, and should be objective. Apart from being the most vital part, it is also generally the lengthiest section of any research paper, unless a methodology paper for that research is published separately.[1] While writing this section, the authors must ensure that it is crisp, concise and complete in every aspect to address all the queries pertaining to the study like 'who', 'where', 'when', 'what' and 'how'.[2]

GENERAL CONCEPTS, FRAMEWORK AND WRITING STYLE

This section, while being written for a research paper, must present exactly the same information as in 'Material and Methods' section of the study protocol or thesis. Thus, it must be meticulously built up, giving the maximum time and thought, while preparing the research protocol. A precise and objectively written methods section in a protocol will save time and efforts, as often it can be simply replicated in the final manuscript. The main difference between the two would be of the tense, i.e. future tense in the protocol and past tense in the paper.[3] Often, authors need to truncate the 'Methods' in the final paper so as to match with the Results that are presented for that particular paper, and also to avoid unnecessary details keeping in mind the recommended word count limits of the journal. Describe methodology in such a manner that it has all the information (including references) reader needs to replicate the study/experiment without access to the detailed study protocol.[4]

First published as Arora SK, Shah D. Writing Methods: How to write what you did? Indian Pediatr. 2016; 53:335–40.

The authors must ensure fluency while writing the methods. A common mistake that makes the write-up appear disjointed is use of both passive as well as active voice within the same section. For example, *"The participants were recruited from the outpatient department of the xyz hospital. We collected blood samples from all the patients. The samples were stored at –80°C. We analyzed the serum antibody levels of the stored samples after all the patients were recruited."*

One must stick to either the point of view of the experiment (passive voice) or the point of view of the experimenter (active voice) throughout a paragraph to ensure coherence.[5] Though style manuals prefer the active voice for medical and scientific writing, the passive voice has its place in the Methods section. Hence, you can alternatively modify the above statement in either of the following ways:

"We recruited participants from the outpatient department of the xyz hospital. We collected blood samples from all the patients, and stored the samples at –80°C. We analyzed the serum antibody levels of the stored samples after recruiting all the patients."

Or

"The participants were recruited from the outpatient department of the xyz hospital. Blood samples from all the patients were collected, and stored at –80°C. The serum antibody levels of the stored samples were analyzed after all the patients were recruited."

You can present the methodology in a structured or an unstructured format, depending on the journal's requirements. Structured representation makes it more clear and objective for easier comparability between different studies. Unstructured format, though lacks objectivity, may be a better option for some types of studies like descriptive and qualitative studies.[6] Print journals often prefer unstructured format to save space. Irrespective of the framework, the components of this section essentially remain the same, and have been provided in Box 4.1 as a checklist.

Study Setting, Duration and Design

Study setting description involves mentioning the place where the study was conducted which may be more than one in case of multicentric studies. Specify the place from where the patients/samples/clinical records were recruited/obtained like out-patient

---

Box 4.1 | **Checklist for methods section**

- Study setting and place
- Duration
- Study design
- Ethical considerations
- Consent and assent
- Funding information
- Patient confidentiality
- Trial registration details (if any)
- Follow standard reporting guidelines
- Participant selection and sampling
  - Definition of participants (cases and controls)
  - Sampling technique used
  - Inclusion criteria
  - Exclusion criteria
- Method of randomization (for controlled studies), including group allocation details
- Blinding details (if applicable)
- Exact procedure/Intervention (with references)
- Primary and secondary outcome variables
- Sample size calculation
- Statistical analysis

---

department/inpatient department/emergency/medical record department, for a hospital-based study. The place of study may be a school, village or district in case of a community-based study. One should also mention all the departments involved in carrying out the study apart from the primary department. The exact duration over which the study was carried out must also be specified, including period of enrolment of participants and their follow-up.

Study design must be specified in the beginning of the methods section. Medical research is broadly classified as primary and secondary. Secondary research involves summarizing the results available from primary research in the form of systematic reviews and meta-analyses. Primary research that involves synthesizing the evidence can broadly be classified into basic medical research, clinical research and epidemiological research; though there is no rigid demarcation in these study areas.[7] The commonly used epidemiological study designs are depicted in Fig. 4.1. Descriptive studies help to generate a hypothesis by simple description of certain population parameters and finding some associations.

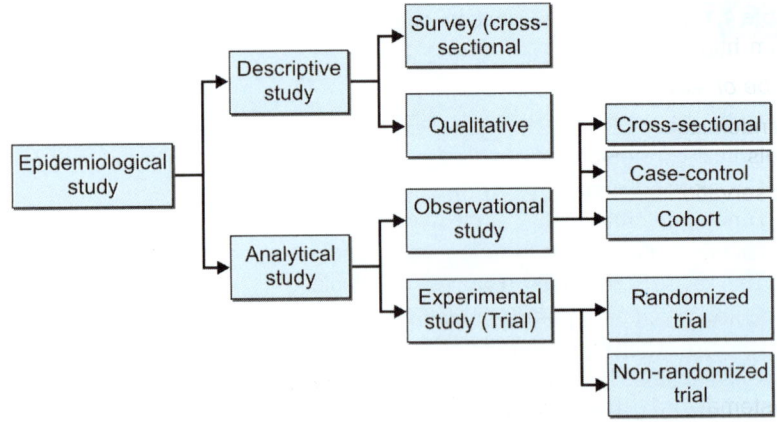

Fig. 4.1: Some common epidemiological study designs

Analytical studies help test such hypothesis to establish causation. Analytical studies can be observational or experimental. The purpose of an observational study is to follow the natural course of events in one or more groups formed on the basis of presence/absence of exposure, risk factor or disease. Whereas in an experimental study, the investigator intentionally manipulates one or more independent variables after controlling the effect of potential confounders and analyzes the results of that intervention.[8] Apart from these, there are certain areas of special research like qualitative research, decision analysis, operations research, health systems research, quality assurance, cost-effectiveness/economic analysis, which are beyond the scope of this article.

The exact layout of methods section depends on the type of study design. Nowadays, majority of the journals recommend the authors to adhere to the respective reporting guidelines for different types of studies to promote transparent, accurate and good quality reporting (Table 4.1).[4,9-17] These guidelines help the authors elaborate the study in detail that makes the evaluation and analysis of medical literature easy for not only the editors and reviewers, but also for the readers and researchers.[4]

Ethical Considerations, Confidentiality, Clinical Trial Registry and Funding Information

All the clinical studies involving human subjects need ethical clearance from the local/institutional ethical committee/review

Table 4.1: Reporting guidelines for different types of study designs (available from http://www.equator-network.org/)[9-17]

Type of study	Reporting guidelines	Website
Randomized trials	**CONSORT** (**CON**solidated Standards of **R**eporting **T**rials)	www.consort-statement.org
Observational	**STROBE** (**St**rengthening the **R**eporting of **Ob**servational **S**tudies in **E**pidemiology)	http://strobe-statement.org/
	RECORD (**Re**porting of studies **C**onducted using **O**bservational **R**outinely-collected health data)[10]	http://www.equator-network.org/reporting-guidelines/record/
Systematic review and meta-analysis	**PRISMA** (**P**referred **R**eporting **I**tems for **S**ystematic **R**eviews and **M**eta-**A**nalyses)	http://prisma-statement.org/
Case reports	**CARE** (**CA**se **RE**ports)	http://www.care-statement.org/
Qualitative research	**SRQR** (**S**tandards of **R**eporting **Q**ualitative **R**esearch)[11]	http://www.equator-network.org/reporting-guidelines/srqr/
	ENTREQ (**EN**hancing **T**ransparency in **RE**porting the synthesis of **Q**ualitative research)[12]	http://www.equator-network.org/reporting-guidelines/entreq/
	COREQ (**CO**nsolidated criteria for **RE**porting **Q**ualitative research)[13]	http://www.equator-network.org/reporting-guidelines/coreq/
Diagnostic/prognostic studies	**STARD** 2015 (**STA**ndards for **R**eporting **D**iagnostic accuracy studies)[14]	www.stard-statement.org/
	TRIPOD (**T**ransparent **R**eporting of a multivariable prediction model for **I**ndividual **P**rognosis or **D**iagnosis)[15]	http://www.equator-network.org/reporting-guidelines/tripod-statement/
Quality improvement studies	**SQUIRE** (**S**tandards for **Q**uality **I**mprovement **R**eporting **E**xcellence)	http://www.squire-statement.org/
Economic evaluations	**CHEERS** (**C**onsolidated **H**ealth **E**conomic **E**valuation **R**eporting **S**tandards)[16]	http://www.equator-network.org/reporting-guidelines/cheers/
Study protocols	SPIRIT (**S**tandard **P**rotocol **I**tems: **R**ecommendations for **I**nterventional **T**rials)[17]	http://www.spirit-statement.org/

board before initiation of recruitment of subjects. The same is applicable for animal studies as well. A declaration about the ethical clearance is mandatory in virtually every medical journal specifying the authority from where it has been obtained. The study must comply with the principles of the Declaration of Helsinki[4]—a set of ethical principles regarding experimentation involving human subjects developed for the medical community by the World Medical Association (WMA).[18] It was first adopted in 1964 and has undergone several revisions with the latest one in 2013. The ICMR guidelines on research on human subjects (available from *http://icmr.nic.in/ethical_guidelines.pdf*) can also be utilized for this purpose.[19]

The authors need to maintain patient confidentiality; hence should refrain from using patients' names, initials, or hospital numbers, especially in illustrative material. One must specify the details regarding obtaining informed consent from the participants/guardians for inclusion in the study and publication of clinical details or/and clinical photographs. Assent must be taken from all the subjects more than 7 years of age and the same must be mentioned in the manuscript.

Most of the medical journals recommend that all clinical trials involving human subjects should be registered in a public trial registry before the onset of patient enrolment, and hence also publish the trial registration number in the manuscript at the end of abstract.[4] Clinical trial registration helps in preventing duplication of research activities, and attempts to prevent selective reporting of research outcomes. One may access these registries to acquire knowledge regarding current research going on in a particular field. Due to these reasons, International Committee of Medical Journal Editors (ICMJE) encourages registration of the studies with non-trial research designs as well, even though it is not mandatory. All the clinical trials being carried out in India need to be registered in the Clinical Trials Registry of India (*www.ctri.in*) which is hosted by the Indian Council of Medical Research. Alternatively, researchers may register their trials in one of the following trial registries: *http://www.actr.org.au*; *http://www.clinical trials.gov*; *http://isrctn.org*; *http://www.trialregister.nl/trialreg/ index.asp*; and *http://www.umin.ac.jp/ctr*. Funding information may be specified in the methodology section, though majority of the journals require a separate declaration regarding funding to be furnished during manuscript submission.

Population, Sample and Participant Characteristics

Authors must describe the population from which they have chosen the participants of the study. A description of how the participants were selected, i.e. the type of sampling technique used (e.g. simple random, stratified, cluster, convenience) is also important. Subjects for the research may be patients, laboratory samples, animals, hospital records, etc. Similarly, for a systematic review or meta-analysis, the subjects will be clinical studies like randomized controlled trials (RCTs). In case of trials, provide details regarding process of randomization and allocation of subjects to the different groups. Specify the methods of allocation concealment and blinding wherever applicable. One should avoid labeling the different groups with alphabets (groups A, B) or numbers (groups 1, 2); rather stick to names, e.g. immunized group and unimmunized group, to minimize confusion to the readers.[20]

In comparative studies like case-control studies and controlled trials, there is a group of subjects that does not receive an intervention, receives a placebo or receives a different intervention. It is equally essential to describe the comparison or the control group characteristics as the main study group. The manuscript must incorporate the criteria for selection of controls and the methodology used for allocation of groups that was used to ensure comparability. The authors must give criteria used for matching, the method of randomization (simple, stratified, block, etc.), and technique of allocation concealment and blinding, wherever applicable.[2]

Specify the detailed criteria (e.g. age, sex, well-defined disease condition) which make the participant eligible to be included in the study. Studies usually also have some exclusion criteria–parameters that make a subject ineligible to be a part of the study even after fulfilling the inclusion criteria. The exclusion criteria are generally outlaid to avoid bias or due to some feasibility or ethical issues. There should not be any overlap in the inclusion and exclusion criteria. In other words, the exclusion criteria can be applied only on the subjects who already fulfill the inclusion criteria. For example, in a study on prevalence of celiac disease in adolescents with anemia, inclusion criteria could be:

"All adolescents (age 10–19 years) residing in a particular community, and having hemoglobin values below the cut-offs (Hb<12g/dL in 10–19 y girls and 10–14 y boys and Hb<13g/dL in 15–19 y boys) [Reference]."

Exclusion criteria for this study could be:

"Those having received a blood transfusion or hematinics in preceding 4 weeks, those with acute illness (e.g. fever, diarrhea, respiratory tract infection) or known chronic disease (e.g. chronic liver disease, chronic kidney disease, thalassemia)."

Take care not to write: Age <10 years or >19 years or Hb >13 g/dL as exclusion criteria as these subjects already do not fulfill the inclusion criteria.

Procedure, Intervention and Outcome Measures

The procedure may be just recording an observation, recording the response to a questionnaire, carrying out a diagnostic test or doing an intervention which may be preventive or therapeutic. Mention the exact details of the intervention applied to the participants in case of trials; description of the drug, device or educational program being tested, including the exact dosage, formulation, schedule and duration. For trials as well as for observational studies, discuss about the process of collecting information and data for analysis. Give description of the variables analyzed, technique and the instruments used in the study. Give the reference if the technique used in the study has been published previously or is a well-established, standardized one. If that is not the case, ensure to describe it well with the exact temporal sequence. Similarly, one must give the manufacturer's name and place in parenthesis if a novel apparatus has been used. For example,

"Fasting blood sample was collected to measure HbA1c by HPLC (BIO-RAD Germany) and lipid profile (enzymatic method). The GE-Lunar DPX Pro (GE Healthcare, Wisconsin, USA) was used to measure body composition."[21]

Outcome measures or study end points are the parameters which will fulfill the objectives of the study and are classified as primary and secondary. Primary outcome, which is generally single, is the parameter on which the study hypothesis is based and is the main objective of research. The other outcomes of interest, which may be more than one, are designated as secondary outcomes. These all should be clearly defined while writing a manuscript. For example,

"The primary outcome was CPAP failure, defined as need for intubation and mechanical ventilation within 72 hours of initiation of respiratory

support… . The secondary outcomes related to respiratory support were duration of CPAP support, duration of supplementary oxygen requirement, maximal flow, PEEP and oxygen requirement, incidence of air leaks and Broncho-pulmonary dysplasia. Other outcomes included incidence of patent ductus arteriosus, intraventricular hemorrhage…".[22]

Statistical Analysis

The data management strategy and statistical analysis technique used for a study must always be provided in sufficient detail, to the extent that any skilled person having access to the original data set is able to reproduce the results. The manuscript must include an account of sample size calculation with its justification as well as literature citation as appropriate. The computer software used for data analysis should also be mentioned. The authors should avoid using generalized statements and must write statements specific to that study parameters and outcome variables. The statistical tests and the comparisons must be specified. Common statistical methods may just be mentioned but advanced or unusual methods must be described or cited with an appropriate reference. Description of statistical analysis of the primary outcome must precede that of secondary outcome(s).[20]

WHAT NOT TO WRITE IN METHODS

Methodology must be described in a complete but concise manner avoiding any unnecessary detail that is irrelevant to the readers. Methods section should include only that information that was available at the time of planning of the study, whereas any information that was collected while carrying out the study should be a part of the Results.[4] Avoid giving any explanatory information in this section like background and rationale for using a particular methodology for a particular study that may be covered under the section on discussion. Methods section must include only the proposed sample size and not what was actually achieved. The account of the subjects who were selected by sampling till the ones who were eventually analyzed, including details like refusal to give consent, exclusion based on exclusion criteria must also be covered under the Results and not Methods.[3]

To conclude, the section on methods is the foundation stone of any research being planned or written; hence it must be clear and elaborate. It should also be sufficiently described for easy reproducibility as JC Jones said, "*Methodology should not be a fixed track to a fixed destination but a conversation about everything that could be made to happen.*" Laying the strong foundation of a robust Methods section will pave the way for smooth writing of the rest of the paper. The next chapter will guide the readers about the intricacies of presenting your results.

REFERENCES

1. Dewan P, Gupta P. Writing the title, abstract and introduction: looks matter! Indian Pediatr. 2016; 53:235–41.
2. Azevedoa LF, Canário-Almeidaa F, Almeida Fonsecaa J, Costa-Pereiraa A, Winckb JC, Hespanholb V. How to write a scientific paper–writing the methods section. Rev Port Pneumol. 2011; 17:232–8.
3. Shah D. Material and methods: How will I do it? *In*: Gupta P, Singh N, editors. How To Write the Thesis and Thesis Protocol. New Delhi: Jaypee Brothers; 2014. p. 75–82.
4. Recommendations for the Conduct, Reporting, Editing, and Publication of Scholarly Work in Medical Journals. Available from: *www.icmje.org/recommendations*/Accessed February 25, 2016.
5. Kallestinova ED. How to write your first research paper. Yale J Biol Med. 2011; 84:181–90.
6. Maxwell JA. Methods: what will you actually do? *In*: Maxwell JA, editor. Qualitative Research Design: An Interactive Approach, 2nd ed. Thousand Oaks, CA: Sage; 2005. p. 79–104.
7. Röhrig B, du Prel JB, Wachtlin D, Blettner M. Types of study in medical research. DtschArztebl Int. 2009; 106:262–8.
8. Porta M, Greenland S, Last JM, editors. A Dictionary of Epidemiology. 5th ed. New York: Oxford University Press; 2008.
9. EQUATOR Network. Oxford: EQUATOR Network. Available from: *www.equator-network.org/reporting-guidelines/*. Accessed February 25, 2016.
10. Benchimol EI, Smeeth L, Guttmann A, Harron K, Moher D, Petersen I, et al; RECORD Working Committee. The REporting of studies Conducted using Observational Routinely-collected health Data (RECORD) Statement. PLoS Med. 2015; 12:e1001885.
11. O'Brien BC, Harris IB, Beckman TJ, Reed DA, Cook DA. Standards for reporting qualitative research: a synthesis of recommendations. Acad Med. 2014; 89:1245–51.
12. Tong A, Flemming K, McInnes E, Oliver S, Craig J. Enhancing transparency in reporting the synthesis of qualitative research: ENTREQ. BMC Med Res Methodol. 2012; 12:181.
13. Tong A, Sainsbury P, Craig J. Consolidated criteria for reporting qualitative research (COREQ): a 32-item checklist for interviews and focus groups. Int J Qual Health Care. 2007; 19:349–57.

14. Bossuyt PM, Reitsma JB, Bruns DE, Gatsonis CA, Glasziou PP, Irwig L, *et al.* For the STARD Group. STARD 2015: An updated list of essential items for reporting diagnostic accuracy studies. BMJ. 2015; 351:h5527.

15. Collins GS, Reitsma JB, Altman DG, Moons KG. Transparent reporting of a multivariable prediction model for individual prognosis or diagnosis (TRIPOD): The TRIPOD statement. Ann Intern Med. 2015; 162:55–63.

16. Husereau D, Drummond M, Petrou S, Carswell C, Moher D, Greenberg D, *et al.* Consolidated Health Economic Evaluation Reporting Standards (CHEERS) statement. Eur J Health Econ. 2013; 14:367–72.

17. Chan A, Tetzlaff JM, Altman DG, Laupacis A, G.tzsche PC, Krleþa-Jeriã K, *et al.* SPIRIT 2013 Statement: Defining Standard Protocol Items for Clinical Trials. Ann Intern Med. 2013; 158:200–7.

18. World Medical Association. World Medical Association Declaration of Helsinki: Ethical principles for medical research involving human subjects. JAMA. 2013; 310:2191–4.

19. Ethical guidelines for biomedical research on human participants. New Delhi: Indian Council of Medical Research; 2006. Available from: *http:// icmr.nic.in/ethical_guidelines.pdf.* Accessed March 17, 2016.

20. Wenzel V, Dünser MV, Lindner KH. A step by step guide to writing a scientific manuscript. Available from: *www.aaeditor.org/StepByStepGuide.pdf.* Accessed February 23, 2016.

21. Parthasarthy L, Chiplonkar S, Khadilkar V, Khadilkar A. Association between metabolic control and lipid parameters in Indian children with type 1 diabetes. Indian Pediatr. 2016; 53:39–41.

22. Goel S, Mondkar J, Panchal H, Hegde D, Utture A, Manerkar S. Nasal mask versus nasal prongs for delivering nasal continuous positive airway pressure in preterm infants with respiratory distress: a randomized controlled trial. Indian Pediatr. 2015; 52:1035–40.

PEARLS OF WISDOM

Preparing and Submitting Photographs/Images/Figures

These are used to document clinical presentation of patients; medical/surgical procedures or specimens; findings from investigations such as radiographs, ultrasonography, computed tomography (CT) or magnetic resonance imaging (MRI); and pathology or microbiology images captured from the microscope. Following points should be kept in mind while using the image:

- Use photographs only if they show something essential to your point, something that cannot be conveyed by text.
- Photographs should be of good quality, not distorted; and labelled properly.
- Proper written consent should be obtained beforehand if clinical photographs of the patient are to be included.
- All attempts should be made to protect the identity of the person by masking using permanent method of editing.
- All figures should be cited in the main text of your article.
- The pictures should also have a legend describing what is being portrayed.
- Do not use any abbreviations in the legends of photographs. If any abbreviation is used to describe findings on the picture, expand it in the footnote or legend.
- Use arrows (using permanent image editing) to highlight important findings.
- No specific feature within the image should be introduced, moved, obscured, removed or enhanced.

Shalu Jain and Dheeraj Shah

Writing the Results

•Aparna Mukherjee •Rakesh Lodha

The culmination of all the efforts put in conducting a study is in obtaining the results of the study; hence, while publishing the study one should be very attentive in writing the results section. In this book on reporting of research, you have already read chapters on writing the Introduction and the Methods sections[1,2]. We herein provide pointers to writing the Results section, which is the cornerstone of a research paper where one can highlight the achievements of the study, it is here that one actually presents the painstakingly collected and analyzed data.

The skill of presenting the findings of one's study clearly and logically, so that it is easily understandable by the reader (who has never been as involved with the study details as the researcher) improves with practice. Many good studies are marred by results that are described in a haphazard, directionless and over-elaborative manner.

In this section, in addition to the text, tables and figures are also used to communicate the content. When well-written, this section can give a lucid picture of what the study is trying to say. Data obtained from the study, when presented coherently, inspires the readers' confidence in the authenticity and robustness of the study. It should be a step-by-step approach, taking the readers along the plan of analysis that was described in the methods section. At the end of this section, the readers should be able to draw their own conclusions about the study findings.

WHAT SHOULD BE INCLUDED IN THE RESULTS SECTION?

Results section is a platform to narrate the observations; no attempt should be made to explain the findings–interpretations should be

First published as Mukherjee A, Lodha R. Writing the results Indian Pediatr. 2016; 53: 409–15.

left for the discussion section. Similarly, details of statistical tests, definitions, and plan of study should be included in the methods section rather than the results section. The guidelines for authors as per the targeted journal should be thoroughly read and instructions followed. Results section is for reporting the findings of your own study and not for any comparison with other studies; however, in systematic reviews and meta-analysis, various included studies' pertinent results are presented. Do not give any data which is not your own and for which a reference has to be cited.

A large amount of data may be generated in a study; however, it is not wise to include all the data together. One has to be careful as to how much of the data are presented. Too much of information might obscure the pertinent findings, whereas too little might render the study incomprehensible and unreliable.

GENERAL STYLE OF WRITING

The style of writing should be fluent and uncluttered, without unnecessary use of adjectives and adverbs. Use past tense when describing the results as all events being reported while publishing the study has happened historically[3]. Write with clarity and brevity, as shown in the following example:

a. *It is clearly seen that average weight of children in group A is markedly higher than that of group B.*
b. *The mean (SD) weight of children in group A [16.5 (2.1) kg] was higher as compared to that in group B [10.3 (1.2) kg], P = 0.04.*

Sentence B is the more appropriate way of expressing the result.

ORGANIZING THE RESULTS TO MAKE THEM MORE MEANINGFUL

The pattern of describing the results may differ in its minutiae according to the study type and the journal. However, the broad outlines are essentially the same. Qualitative studies have a slightly different approach than quantitative studies; we will concentrate on the latter in this paper.

Where to Start–Details of Study Subjects/Outline of Study?

One should start with data regarding number of patients/subjects enrolled. Numbers of potential subjects screened, numbers excluded for not meeting study criteria, numbers randomized, and numbers finally being analyzed along with numbers of trial deviates and lost to follow up should be clearly mentioned; use of a flow diagram may be useful. Following is an example to highlight these aspects:

"A total of 300 [175 (58.3%) girls] children with tuberculosis, aged 6 months to 15 years of age, were enrolled after screening 800 children attending the pediatric outpatient department of our hospital."

Thereafter, a description of study subjects, the demographic and clinical characteristics can be presented in the text or in a table. Most editors/reviewers do not want the P value to be written when presenting the baseline demographic characteristics of the intervention and control arms of randomized controlled trials. Number of participants with missing data for each variable of interest should be indicated. Data regarding exposures and potential confounding factors should be mentioned here.

Which Reporting Guideline to Follow?

The various guidelines for reporting studies as well as the instructions to the authors by the journals should be closely followed. Details of these guidelines can be obtained from the link: *http://www.equator-network.org/*

The STROBE guidelines should be referred to in cases of observational studies—both case control and cohort studies[4]. It is advisable to include a flow diagram indicating the study outline. In case of cohort studies and cross-sectional studies, number or the summary measure of outcome events should be mentioned. In case control studies, the numbers in each exposure category or summary measure thereof should be reported.

In case of a randomized control trial, CONSORT statement for reporting should be followed and a study flow diagram is essential[5]. Dates defining the periods of study recruitment and follow-up should be mentioned. If the trial is stopped prematurely, reasons for doing so should be clearly specified. All adverse events, even if unrelated to the study intervention should be mentioned in each group.

While reporting diagnostic studies, the STARD guidelines are to be followed[6]. A flow diagram can be included to depict the enrollment of participants. The distribution of severity of disease in participants with the target condition and alternate diagnosis in those without the target condition has to be mentioned.

A 2 × 2 table should be provided for the test result under study and the reference or gold standard test. Diagnostic accuracy should be stated along with the 95% confidence interval. Any adverse events encountered while performing the test being studied and the reference standard should be reported.

Presenting the Outcomes: The Sequence

It is important to have a proper sequence for the presentation of the observations of the study. Often, the authors present the observations where the p value is the lowest/significant. At times, the presentation includes all the data collected with the main outcomes coming somewhere in between. It becomes difficult for the reader to follow such a haphazard arrangement.

Addressing the Objectives

The primary objective of the study should be dealt with in the beginning. The data should be so stated that the primary question asked is answered clearly, using the statistical tests as described in the methodology section. Refrain from presenting raw data. Thereafter, the secondary outcomes, if any, should be presented.

Remaining data should follow the order as described in the methodology section. It is advisable to go from the simpler to more complicated results. Subgroup analysis should come in the last. Do not describe any finding in the result section for which the methodology and statistical analysis plan has not been stated in the methods section. On the other hand, present all relevant results as mentioned in the methodology section; do not exclude the observations from any test or investigation which has been mentioned in the methodology section. In case some post hoc analyses are performed, the same should be stated explicitly.

Results section can be divided into subheadings according to the objectives studied and analysis done, in order to increase the clarity; however, the same should be done as per the journal's format.

IMPORTANT ISSUES WHILE WRITING RESULTS SECTION

Although you would like to write a comprehensive results section and include all your study findings; frequently we miss some essential attributes of this section. Some of these are given in Box 5.1.

Box 5.1 **Check these before finalizing the results section**

Decimal points
- Usually one or two places after decimal point are sufficient.
- Be consistent with the format.

Importance of P value
- Write the actual P value.
- Never state a P value as 0.000.

Choose your words carefully
- Be cautious while using the word 'significant'.

Confounders
- Make clear which confounders were adjusted for.

Negative results
- Always report the negative findings as well.

Text – table dichotomy
- Avoid repetition between text and tables.

Decimal points

Usually the computer programs that we use for statistical analysis will return results with a lot of digits after the decimal point like 3.4562789. Do not report as such; round off to the decimal point which reflects the sensitivity of your measuring instrument or assay. For example, if the birth-weight is measured in grams, then the mean value for the weight should have only one decimal place rather than 2 or more decimal places. Usually one or two places after decimal point are sufficient. However, one should be consistent with the format which should be as per the journal's requirement.

Importance of P value

The significance of the statistical tests applied is usually presented as P values; the actual P value should be written instead of just stating <0.05. Never state a P value as 0.000—this does not make sense even if the statistical program does return such values; such values should be stated as <0.0001. Many reviewers and journals nowadays prefer the 95% confidence interval over the P values[5].

Choose your words carefully

One should be cautious while using the word 'significant'—in the results section, it denotes that the difference is statistically significant and not by chance, i.e. P<0.05. Do not use the word if you find any difference which is not statistically significant.

However, do present the statistically non-significant data as well. Defer from using words like 'about' or 'approximately' while describing the results.

Confounders

In scenarios where confounders are expected to alter the results, first give the unadjusted estimates and then the confounder-adjusted estimates and their precision (e.g., 95% confidence interval). Make clear which confounders were adjusted for and why they were included; the process for the same should be described in the methods section.

Negative results

The researcher is usually biased towards the positive findings of his or her study. However, the negative results obtained from the studies are equally important; they may help to prove or disprove many proposed hypothesis. Remember to always report the negative findings as well. Not reporting the negative results is unethical and reduces the authenticity of the data.

Text–table dichotomy

Text and tables should be complementary to each other, not repetitive. Some salient features informing the reader as to what is described in the tables can go in the text; rest can be depicted in the table, for example:

"A total of 300 [175 (58.3%) girls] children with tuberculosis, aged 6 months to 15 years of age, were enrolled. Mean (SD) age of the children was 110 (15.3) months. Table … shows the baseline characteristics of the enrolled children."

Tables

Tables are a good way of presenting large amount of complicated data in a structured fashion. A lot can be communicated through tables, but care should be taken so that the tables are simple and comprehensible. The tables and text should not contain the same detailed information, this is considered redundant. Only the key message of the table can be concisely described in the text. The text should always be linked with the table by referencing the table sequentially. Depending on the journal, tables may be presented sequentially at the end of the manuscript after the 'References', or located within the text of your results section[7].

In a study with the objective to document the weight gain at the end of 6 months of anti-tubercular therapy, the result can be written as:

"At 6 months, median (IQR) weight gain was 3.5 (3.1, 4.7) kg and WAZ was 0.65 (0.59, 0.79). Change in WAZ was assessed at 2 and 6 months and are described in Table XYZ."

Number of Tables

Number of tables to be included is primarily decided by the requirement of the publishing journal. Usually a maximum of 3-4 tables should be sent. Remaining data may be submitted as supplementary material in case of online publications.

Requisites of a Good Table

Each table should be complete in itself; one should be able to read the table without taking help from the text. The essential components of the table include a heading or legend, row and column headers and footnotes. Heading should be short, specific, descriptive, stating the key message to be enumerated in the table. Do not use abbreviations in the heading. Each row and column header should be able to explain what the row/column contains. The footnote should contain the abbreviations mentioned in the table and any other explanatory notes required. Always mention what summary statistic is being presented like N (%), mean (SD) or median (IQR) and also the unit of measurement. This can be mentioned in the appropriate column/row header or in the footnote (Table 5.1). Indicate number of participants with missing data for each variable of interest. Do not just give percentages if the denominator is less than 100, rather give the actual observed values.

When comparisons are being made, the data should preferably be presented side by side.

Do not combine disparate variables in the same table. If a table becomes too long, it is better to split into two for better understandability.

Figures

Figure give a visual key which is usually appealing to the reader. Decide which data can be best presented in the form of figure; it is usually the important information which is presented in this format–something which you want the readers to easily understand and retain. Figures can be in color or grey tone. Almost all journals

Table 5.1: Factors associated with treatment outcome in enrolled children

Characteristics	Treatment success n = 80	Poor outcome n = 45	P value
Age (mo), mean (SD)	99.1 (43.1)	113.7 (42.3)	0.2
ender, n (%)			
Male	43 (53.7)	20 (50)	0.9
TST positivity, n (%)	58 (79.4)	27 (75.0)	0.6
Contact with TB, n (%)	21 (38.6)	18 (28.2)	0.2
BCG vaccinated, n (%)	45 (58.9)	20 (44.2)	0.1
Diagnosis, n (%)			
Pulmonary	51 (63.7)	29 (64.4)	0.5
Extra-pulmonary	29 (36.3)	16 (35.6)	
Weight for age	−1.7	−1.8	0.4
Z score at the start of therapy, median (IQR)	(−2.3,−0.9)	(−2.4, −1.2)	

AFB: acid fast bacillus; TB: tuberculosis; TST: tuberculin skin test;
BCG: Bacillus Calmette Guerin; ATT: anti-tubercular therapy;
IQR: interquartile range

charge extra for colored figures. Again the number of figures to be included is to be decided as per the requirement of the publishing journal.

Figures may range from simple line diagrams to scatter plots to radiographs/images. Figure should also be complete in itself with an informative heading with no abbreviations. Legends, data labels, axis titles, etc. should all be complete[7].

The text, table and figure should be complimentary to each other and not mere duplication of data. Figures should be cited in the text and should be numbered in the order of reference in the text.

All figures and pictures should be submitted as separate files in the form of image file (.jpg, .ppt, .gif, .tif or .bmp) with minimum resolution of 300 dpi to ensure good print quality; the authors should refer to the target journal's instructions to choose the correct file type and the resolution.

Types of Figures/Illustrations

1. *Photographic images:* These images are used to document observations such as clinical photographs of patients; data of imaging investigations such as radiographs, ultrasonography images, CT scan/ MRI scan images, radionuclide studies;

intraoperative findings; surgical specimen; pathology images-cytopathology, histopathology, special stains, immunohisto-chemistry, etc.; laboratory investigations such as PCR results, gel/blot images; tracing of investigations such as ECG, EEG, EMG, etc.

Images should be of good quality, maintaining the original proportions (not unnaturally distorted), cropped to delete unnecessary details and labeled properly. If you plan to include photographs of your patient, proper written consent should be obtained beforehand. Patient confidentiality is of paramount importance and all attempts should be made to protect the identity of the person like covering the eyes. The pictures should also have a legend describing what is being portrayed. At times, it is useful to combine many images in a single figure, e.g. CT scan/MRI scan images; each of the images should be identified separately. Use of arrows or other markers may be helpful to highlight important findings.

2. *Graphs/data chart:* Data charts can effectively summarize numerical data for better presentation.

Choosing the right type of graph for your data is critical. The right graph depends on a number of factors like the type of data (continuous or categorical), the number of groups or variables involved and the intent of creating the graph. When one wants to demonstrate the composition or break up of a data set or groups within the data set, one can use pie charts or stacked bar charts. In order to show comparison between data, you can use bar charts or line plots. Line plots can be used to depict time trends. Distribution of data can be demonstrated by histograms, scatter plots. Overlapping of data can be picturized by Venn diagrams. The relationship or correlation between two variables can be demonstrated by scatter plots. Commonly used data charts are:

a. *Pie chart:* Pie chart shows classes or groups of data in proportion to the whole data set. They are usually beneficial to depict large data sets, e.g. epidemiological surveys (Fig. 5.1). These are best used when the number of classes/groups are 3–10. One should avoid using the pie chart where there are only 2 groups, e.g. gender.

b. *Bar charts:* Bar charts may be horizontal or vertical. The height or length of the bars represents the measurement. By this method the same variable can be compared across groups or time points (Fig. 5.2). Stacked bars can also be made to make intra-group comparisons like in males and females or to show the composition of each group. It is better not to compare two variables in the same chart if values of one variable

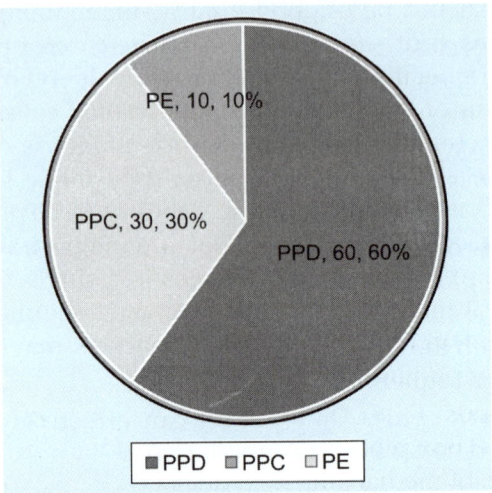

PPD: Progressive pulmonary disease; PPC: Primary pulmonary complex; PE: Pleural effusion.

Fig. 5.1: Pie chart depicting the diagnosis of enrolled chilren with tuberculosis.

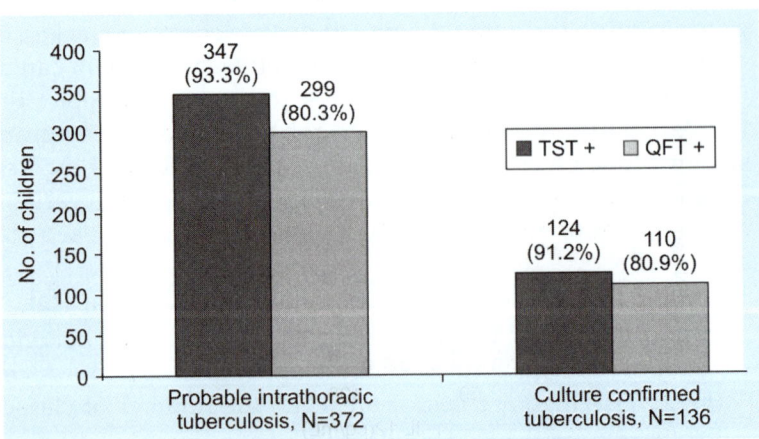

TST: Tuberculin skin test; QFT: Quanti FERON gold in tube test.

Fig. 5.2: Comparison of tuberculin skin test and quantiferon gold in-tube test in children with probable and culture confirmed intrathoracic tuberculosis.

overshadows or dwarfs the other. Also it is prudent to avoid clubbing too many variables or categories in the same chart—this makes the chart unreadable. If there are more than 5 groups to be compared, better to use horizontal bar charts.

c. *Histogram:* In a histogram, the entire range of a continuous variable is divided consecutively into non-overlapping groups known as class intervals. The height of the vertical rectangles for each class interval represents the frequency or density of the variable, depending upon whether the class intervals are of equal or unequal width.

d. *Scatter plots:* Scatter plots can be used to present measurements on two or more variables that are related; the values of the variables on the y-axis are dependent on the values of the variable plotted along the x-axis (Fig. 5.3).

e. *Line plots:* Line plots are similar in some ways to the scatter plots, with the condition that the values of the x variable have their own sequence (Fig. 5.4). Line plots can be used to depict time trends.

f. *Box plots:* Box plots are used to depict the numerical data in the form of median and interquartile range (which forms the box); sometimes whiskers are added which may denote the maximum and minimum value. Outliers can also be depicted in the box plot (Fig. 5.5).

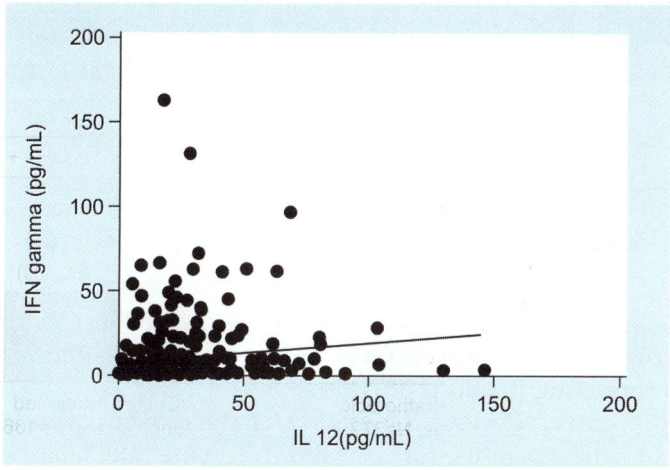

IFN-gamma: interferon gamma; IL 12: interleukin-12

Fig. 5.3: Scatter plot showing the relation between serum interferon-gamma and interlaukin-12 level in 150 neonates at baseline.

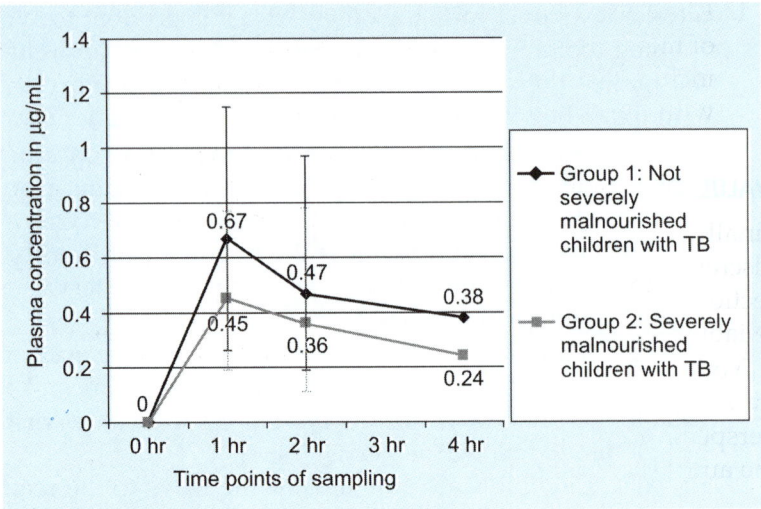

Fig. 5.4: Comparison of plasma concentrations of isoniazid in children with tuberculosis, with or without severe malnutrition.

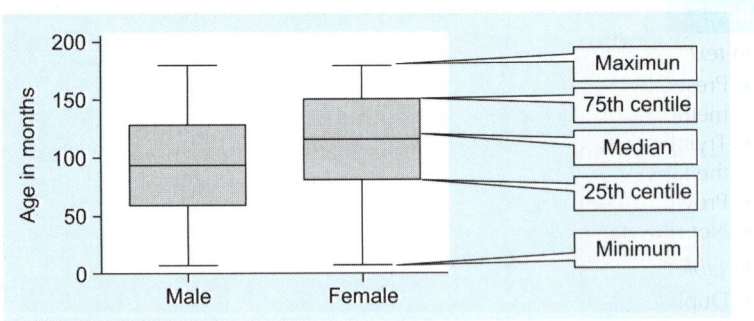

Fig. 5.5: Summary of age of 402 enrolled children with pulmonary tuberculosis.

g. *Venn diagram:* A Venn diagram is a type of chart that shows how different data sets relate to or overlap each other through intersecting portions of circles.

h. *ROC curves:* Receiver Operating Characteristic curve is a plot of the true positive rate (sensitivity) against the false positive rate (1-specificity) for the different possible cut-offs of a diagnostic test. The area under the curve is a measure of accuracy of the test under question.

i. *Forest plot:* Forest plot is a graphical presentation of the results of meta-analysis, where the individual estimated effect of the included studies with the same objective is portrayed along with the cumulative effect.

VALUE OF REVISION

Finally, check and recheck your data. There should not be any discrepancy or inaccurate reporting. Discrepancies within the result section leave a bad impression on the reviewer, and question the reliability of the data.

You may want to get your result (or the whole paper) reviewed by a colleague before submission to any journal—a neutral perspective often brings out many flaws which are not visible to the author himself.

Common Errors

Box 5.2 lists some of the commonly encountered errors in the results section. Enough attention should be paid to avoid these errors.

Box 5.2 **Common errors encountered in results section**

In text
- Presenting the data haphazardly, not following the order mentioned in the methodology section.
- Trying to explain the results, giving one's own interpretation instead of stating the facts.
- Providing too little or too much information.
- Not reporting the negative findings.

In tables
- Duplication of information in text and tables.
- Table not complete in itself.
- Disparate characteristics and comparisons clubbed together in one table.
- Not linking the table with the text with proper reference and in the right chronological order.
- Inaccurate arithmetic numbers do not add up.

In figures
- Graph not plotted to scale.
- Data not properly labeled.
- Omission of proper legends.
- Data not consistent with text.
- Not linking the table with the text with proper reference and in the right chronological order.

SUMMARY

The results section is the platform where the researchers present their data in an informative and lucid manner for the understanding of the readers. No explanation or interpretations are to be presented in this section. Properly labelled tables and figures, chosen according to the data types, add to the value of this section. The sequence of the presentation of results should always follow the order mentioned in the methodology section. The primary objective of the study should be addressed up-front and with utmost clarity. Maintaining the accuracy and authenticity of the data presented is sacrosanct.

REFERENCES

1. Dewan P, Gupta P. Writing the Title, Abstract and Introduction: Looks matter! Indian Pediatr. 2016;53: 235–41.
2. Arora SK, Shah D. Writing Methods: How to write what you did? Indian Pediatr. 2016;53: 335–40.
3. Kallestinova ED. How to write your first research paper. Yale J Biol Med. 2011;84: 181–190.
4. STROBE statement. Available from: *http://www.strobe-statement.org.* Accessed February 12, 2016.
5. The CONSORT Statement. Available from: *http://www.consort-statement.org/.* Accessed February 12, 2016.
6. STARD 2015: An Updated List of Essential Items for Reporting Diagnostic Accuracy Studies. Available from: http://www.equator-network.org/reporting-guidelines/stard/. Accessed February 12, 2016.
7. Aggarwal R, Sahni P. The Results Section. *In:* Aggarwal R, Sahni P (eds). Reporting and Publishing Research in the Biomedical Sciences. 1st ed. National Medical Journal of India, Delhi. 2015. p. 24–44.

6

Reporting Statistics in Biomedical Research Literature: The Numbers Say it All

• *Amir Maroof Khan* • *Siddarth Ramji*

Statistics is the cornerstone of evidence-based quantitative research. Statistical analysis and outputs pave the way for clinical and policy decision-making, thereby impacting the health status of countless individuals across the world. Various studies have shown that statistical reporting is inappropriate and incorrect in biomedical journals[1-3]. Uniformity and transparency in statistical reporting strengthens the validity and reliability of the scientific literature. Most biomedical researchers are not comfortable with statistics and hence the reporting of the statistical output is quite varied, confusing and meaningless to the reader. This article attempts to provide an overview of good practices of reporting statistics in biomedical research literature.

The statistical reporting guidelines and styles being presented in this chapter draw from instructions for authors of *Indian Pediatrics,* International Committee of Medical Journal Editors (ICMJE) guidelines, Enhancing Quality and Transparency of Health Reporting (EQUATOR) guidelines, and current practices observed in the published research literature[4-6].

As different journals have their specific requirements regarding reporting of statistics, it is necessary that the authors go through the instructions for authors and some already published articles of the journals to which they intend to submit their manuscript for publication. However, the guiding principle of reporting statistical analyses is to *"Describe statistical methods with enough detail to enable a knowledgeable reader with access to the original data to verify the reported results[4]."*

First published as Khan AM, Ramji S. Reporting Statistics in Biomedical Research Literature: The Numbers say it all. Indian Pediatr. 2016; 53: 409–14.

REPORT STATISTICS WHEN RELEVANT

"Can we do some statistical analysis and report from this data here?"

This is a common non-specific question which biomedical researchers ask their colleagues who help them in data-analysis. The requirement in this case is to insert some statistical test result in the manuscript, without giving a thought to the fact as to how it would fit in with the research objectives. The question reflects the investigator's lack of clarity and understanding of their research/ study objectives. Had they been clear about their research objectives, the question would have been something like this: "How to compare these means?" Or, "What is the effect of this variable over another one?" Or, "Is there an association between these variables?" The point of serious concern is that even the peer-review process, at times, fail to pointedly enquire about the relevance and the resultant interpretation of the statistical tests applied.

At times, biomedical researchers are not even aware of the purpose of statistics being applied in their research. They just want to have some statistical results, because it is a 'cool' thing to do, or due to their belief that it increases the chances of publication. Some are of the notion that without reporting a 'p-value', a study is considered as irrelevant. An understanding of the conceptual framework of the study is important before embarking on the statistical analysis. The appropriate application of statistical tests would then aid in the meaningful deconstruction and interpretation of the study results. This would also make the discussion and the conclusion sections more meaningful. Statistics is a vital part of biomedical literature, but only if it's relevant. Else it loses its importance. Statistical tests should be applied and reported where it is relevant and not just for the sake of reporting it.

Statistical tests are used in biomedical research broadly for two reasons:

1. *Estimation studies:* In this type of studies, there is no hypothesis statement. The research question is to find out a 'population estimate' from a given 'sample data'[7]. Examples of such estimates include determining the mean weight for age Z-score (WAZ) of under-five children or proportion of underweight births in the newborn population. In these two examples, the mean WAZ score and the proportion are the statistics to be estimated using the sample dataset. In both these cases there is no hypothesis testing involved. Statistical applications for such objectives will

primarily be restricted to reporting of Standard Errors of Means and Proportions and their related confidence intervals (CI), and will be devoid of any p-value. Statistics is employed here to extrapolate the result from the sample data to the population data in quantitative terms. Hence, statistical tests and the accompanying p-values are irrelevant while reporting population estimates.

2. *Inferential studies (Hypothesis testing):* These studies intend to determine an association between two or more variables. These studies usually are designed to test a hypothesis.[7] Case control studies, experimental studies (randomized or non-randomized, with or without control group), typical cohort studies have a null hypothesis at the start and fall in this category. Statistical tests of significance and the accompanying p-values and effect sizes become relevant and should be reported in studies having hypothesis testing as an objective.

In a research manuscript, statistical methods and statistical outputs are reported in the Methods section and the Results section, respectively. What and how you report the statistical methods and outputs in your manuscript will also depend on the journal you are submitting your manuscript to.

Reporting of Statistical Methods in the Methods Section

The statistical methods reported in the methods section will depend on the study design and objective of the study. Box 6.1 presents the key points to be noted while reporting statistical methods in the methodology section of a biomedical research manuscript.

Box 6.1 Statistical information to be included in the methods section

- All the statistical methods employed in the study.
- Any data transformation done for the purpose of statistical analysis.
- Identify any uncommon statistical method applied, with a reference.
- The order of the statistical methods described should follow that of the objectives mentioned.
- Report the statistical tests in the context of the research objectives, and not as generic statements.
- Use appropriate statistical tests for analyzing paired data.
- Consider the type of distribution while selecting and reporting statistical tests.
- Mention the statistical analysis software packages used for data analysis only for complex analysis.

In the case of observational studies, this section should report all methods used, including those for confounder control. The section must address how missing data was addressed (including loss to follow-up in cohort studies), subgroup analysis if any that were planned, and how matching was done for cases and controls (in the case of case-control studies). When reporting randomized trials, the methods used to compare the primary and secondary outcomes and any additional analysis that were planned must also be reported in this section. The details of what to report in the method sections are available in the reporting guidelines for each study design.[6] The statistical tests employed should be mentioned with respect to the variables being analysed, rather than as standalone general statements. Some examples of stating the test used could be *"Categorical data to test for presence of association between failure to regain birth weight and the likely risk factors was analysed using Fishers exact test"*, or *"The strength of association between the factors and failure to regain birth weight among the infants studied was determined using odds ratios and confidence intervals"*[8].

This section must also report methods used to transform raw data prior to analysis such as converting non-normal to normal distribution, collapsing categories in categorical data, etc. While reporting common tests used such as Chi-square, Fisher exact, student t-test, linear regressions, no citations are needed. However, when reporting more complex analysis, the authors must cite the source, which should preferably be a standard textbook[9]. There is no need to report the analytical software used for the basic descriptive analysis. In the case of statistical analysis involving hypothesis testing, reporting the analytical software is useful and very much required. The alpha level used to define statistical significance must also be mentioned in this section. Box 6.1 presents the key points to be noted while reporting statistical methods in the methodology section of a biomedical research manuscript.

Reporting of Statistical Results

The numerical results must be presented keeping the study objectives in mind. Box 6.2 presents the important points to be kept in mind while reporting statistics in the results section of a research manuscript.

> **Box 6.2** | **Reporting statistical outputs in the results section of the manuscript**
>
> - Avoid nontechnical uses of technical terms in statistics.
> - Explicitly state the groups being compared.
> - Report exact P-values, and not just as significant or non-significant.
> - Do not report P-values as 0.000. In such cases, report as $P < 0.001$.
> - Report the effect sizes with their confidence intervals.
> - Usually P-values with effect sizes and the associated CI are sufficient while reporting the statistical test results, unless the journal asks for additional details.
> - Do not use the term 'correlation', a statistical method to assess the relationship between two continuous variables, to describe 'association'.
> - Refer to relevant reporting guidelines for the study design e.g., STROBE, CONSORT.

What must be included?

For descriptive statistics, the point estimate and 95% confidence interval should be reported. For comparative studies, rates, risk, ratios or the mean difference along with their precision such as 95% confidence interval or standard deviation must be reported.

If P-values are included, the actual value up to 1 or 2 decimal spaces may be reported (e.g. $P = 0.2$ or $P = 0.41$); values less than 0.001 should be reported as $P<0.001$. P-values should not be reported as "not significant" or "NS".

When reporting outcome results, first the results of primary outcomes must be reported and later that of the secondary and any other sub-group analysis. Post-hoc analysis that had not been pre-specified must not be reported.

What can be omitted?

Most statistical softwares will churn out a plethora of outputs during analysis. Typically the outputs would include test statistic (e.g., chi-square statistic, t-statistic, F-statistic), p-value, and degrees of freedom; all of these can be omitted including p-values (but some reviewers would insist on the reporting of p-values). In manuscripts reporting randomized controlled trials, p-values should not be reported when comparing baselines variables/characteristics. Similarly, regression analysis outputs would include multiple data outputs which may include coefficients, r^2, standard errors and p-value. It is best to avoid reporting these in the results unless they serve a useful interpretive function. It is also best not to include complex statistical formulas[10].

Study Design-specific Statistical Results

Observational studies: For these, provide unadjusted estimates and, if applicable, confounder-adjusted estimates and their precision (e.g., 95% confidence interval). Make clear which confounders were adjusted for and why they were included. Report category boundaries when continuous variables were categorized[11]. If relevant, consider translating estimates of relative risk into absolute risk for a meaningful time period.

Randomized controlled trials: For these, provide results for each primary and secondary outcome, and the estimated effect size and its precision (such as 95% confidence interval). For binary outcomes, presentation of both absolute and relative effect sizes is recommended[12].

When reporting correlation, identify the correlation being reported—pearson or spearman. Report the 95% CI and the P- value. While reporting regression analysis, it is best to depict in a tabular format.

CONCLUSIONS

Reporting statistics in biomedical research literature has become more transparent, uniform and reliable. The medical researcher should report relevant statistics and provide meaningful interpretations in their research manuscripts. Pvalues should not be reported alone. If at all, they should be reported along with effect sizes and their confidence intervals. The test statistic and degrees of freedom can usually be omitted. Both, the summary statistics and the inferential statistics should be given careful consideration while reporting. Specific statistical tests require specific components to be reported. Various guidelines are now available to aid the medical researcher for reporting methods. It is important that the specific journal guidelines regarding reporting of statistics should be strictly followed when preparing and sending a manuscript for publication.

REFERENCES

1. Hassan S, Yellur R, Subramani P, Adiga P, Gokhale M, Iyer MS, et al. Research design and statistical methods in Indian medical journals: a retrospective survey. PLoS One. 2015;10: e0121268.
2. Jaykaran, Preeti Y. Quality of reporting statistics in two Indian pharmacology journals. J Pharmacol Pharmacother. 2011;2: 85–9.

3. Horton NJ, Switzer SS. Statistical methods in the journal. N Engl J Med. 2005;353: 1977–9.

4. International Committee of Medical Journal Editors. Recommendations for the Conduct, Reporting, Editing and Publication of Scholarly Work in Medical Journals Available from: *http://www.icmje.org/news-and-editorials/ icmje-recommendations_annotated_ dec15.pdf.* Accessed August 18, 2016.

5. Indian Pediatrics. Statistics, Instruction to Authors. Available from: *http:/ /indianpediatrics.net/author1.htm# Statistics.* Accessed July 16, 2016.

6. Enhancing the Quality and Transparency of Health Research (EQUATOR) Guidelines. Available from: *http://www.equator-network.org/* Accessed July 16, 2016.

7. Hypothesis testing and estimation. *In:* Jennifer P, Barton B, Elliot E, editors. Statistics Workbook for Evidence-based Health Care. 1st ed. West Sussex, UK: John Wiley & Sons; 2009.

8. Namiiro FB, Mugalu J, McAdams RM, Ndeezi G. Poor birth weight recovery among low birth weight/preterm infants following hospital discharge in Kampala, Uganda. BMC PregChildbirth. 2012;12:1.

9. Haruhiko F, Yasuo O. A Guideline for Reporting Results of Statistical Analysis in Japanese Journal of Clinical Oncology. Jpn J Clin Oncol 1997; 27: 21–7.

10. Lang TA, Altman DG. Basic statistical reporting for articles published in clinical medical journals: the SAMPL Guidelines. *In:* Smart P, Maisonneuve H, Polderman A (eds). Science Editors' Handbook, European Association of Science Editors, 2013.

11. von Elm E, Altman DG, Egger M, Pocock SJ, Gotzsche PC, Vandenbroucke JP. The Strengthening the Reporting of Observational Studies in Epidemiology (STROBE) Statement: guidelines for reporting observational studies. Ann Intern Med. 2007; 147: 573–7.

12. Schulz KF, Altman DG, Moher D, for the CONSORT Group. CONSORT 2010 Statement: updated guidelines for reporting parallel group randomised trials. Ann Int Med. 2010; 152: 726–32.

Discussion: The Heart of the Paper

• *Arvind Bagga*

The purpose of 'Discussion' section is to interpret the meaning of results, justify their significance and suggest avenues for further research. This section explains how the research questions or hypotheses presented in the Introduction[1] have been addressed by the results[2], and their impact on the understanding of the research problem. Formulating the discussion requires analytic thinking for synthesis and interpretation of findings and defining key messages that emphasize the implications of research[3]. Drafting an articulate discussion requires scholarly confidence and an ability to think and write creatively. For the majority of authors, this section of the research paper is the most challenging!

This chapter endeavors to simplify this seemingly tough task by outlining its key components(Box 7.1) . The discussion is crafted around the following questions: What do the results mean? How true and relevant are the findings? Why are they important or unique? How does this study expand current knowledge? Do they have any clinical implications? Is there a "take home message"?

WRITING DISCUSSION SECTION

Interpret and Highlight Importance of Results

The analogy of an inverted funnel is used to describe the discussion; the flow of information goes from narrow (focused and precise) to broad[4]. Therefore, the opening paragraph should be a clear and succinct answer to the research question that was posed at the end of Introduction[1,5]. Several illustrations highlight the importance of a direct and explicit opening.

First published as Bagga A. Discussion: The heart of the paper Indian Pediatr. 2016; 53:901–04.

> **Box 7.1   Key components of discussion**
>
> - *Opening paragraph:* Summarize what was done and what was found, e.g., "In order to determine the genetic factors implicated in autism, we used next-generation sequencing to examine 44 genes in 108 patients and 100 controls. We found variations involving multiple genes ..."
> - *Limitations and strengths:* Identify and acknowledge the limitations, than to have them pointed out by the reviewers. Discuss concerns of internal and external validity and generalizability of the results. Mention positive attributes (randomized trial, homogeneous population, large cohort, novelty) without 'claiming the first' and sounding pompous.
> - *Subsequent paragraphs:* Explain the key findings, one by one. For each, summarize what was found and explain how it confirms or refutes what is known on the topic. If there are methodological differences or limitations in the present study, discuss how they impact the results.
> - In the next 1-2 paragraphs, focus on the primary outcome. For example, if you found a novel mutation associated with autism, describe the mutation, its population frequency, and properties and function of the translated protein. Discuss what is known about the gene and protein in relation to brain development, and association with other diseases. Scholarly references enhance the discussion.
> - *Concluding paragraph:* Conclude with a sentence each on the main finding(s) of the study and its clinical relevance. Make focused suggestion for research that would further improve the understanding of the disease or result in better therapies.

- In a recent paper in the *NEJM*, Hoberman, et al.[6] begin the discussion as follows: *"Antimicrobial prophylaxis in children with vesicoureteral reflux diagnosed after a first or second urinary tract infection was associated with a halving of the risk of febrile or symptomatic recurrences. Differences between the prophylaxis and placebo groups were apparent early-on, and increased over a two-year period. Children with bladder and bowel dysfunction at baseline and children whose index infection was febrile derived particular benefit from prophylaxis, with reductions in recurrences of approximately 80% and 60%, respectively"*[6].

- In a randomized controlled study in patients with steroid resistant nephrotic syndrome, Gulati, et al.[7] state: *"Results of this trial show that treatment with tacrolimus and alternate-day prednisone was effective and safe in inducing and sustaining remission in patients with newly diagnosed, steroid-resistant nephrotic syndrome"*[7].

- A recently published randomized study in the *NEJM*[8] begins the discussion as follows: *"The results of our trial showed that withholding parenteral nutrition for 1 week in the pediatric ICU was clinically superior to providing early parenteral nutrition; late parenteral nutrition resulted in fewer new infections, a shorter duration of dependency on intensive care, and a shorter hospital stay. The clinical superiority of late parenteral nutrition was shown irrespective of diagnosis, severity of illness, risk of malnutrition, or age of the child"*[8].

Two additional points need emphasis. While the most relevant findings are explained, how the results support the answer to the research question should be defined at the beginning of the discussion. Repetition of background information or results is unnecessary and tedious. It also increases the length of the manuscript and reduces reader interest. Secondly, the significance of the study results may not immediately be as apparent to the readers, as it is to the author who has devoted a significant amount of time to review the literature, design and conduct the study, and analyze the data[5]. The discussion should attempt to ensure that the readers grasp the importance and uniqueness of the study the very first time they read the manuscript. The central message and relevant findings should be expressed convincingly enough not to be overlooked, even with a quick reading of the discussion.

Acknowledge Limitations and Strengths

Authors can engage their readers effectively with a balanced presentation that emphasizes the strengths but acknowledges the limitations. One should consider all possible alternative reasons for the study results, without being prejudiced to consider only those explanations which support the proposed hypothesis. Secondly, while it is relevant to address limitations on study design, methods or patient number, their implication on validity of the results should be stated[3-10]. Common concerns related to internal validity are issues related to study design, measurement and statistical power. Threats to external validity include sample bias and patient/sample characteristics that limit generalizability of findings. The author(s) may present counter-arguments to mitigate the limitations, or mention strengths to end on a positive note[11].

As an example, while accepting limitations, Gulati, et al.[7] state: *"This study …. was not stratified on renal histology, although there was equal distribution of histopathologies in both arms. While the NIH trial was limited to patients with FSGS, we also included patients with minimal change disease. Although not powered for subgroup analysis, we found that therapy.... was effective in patients with minimal change disease and FSGS. The strengths of this adequately sized trial were that randomization, data collection, and analysis were performed centrally. The baseline characteristics were well balanced, and a widely accepted definition of steroid resistance was used. Safety monitoring was ensured and nonresponders were managed appropriately. The results of this study on children with initial and late resistance and major biopsy diagnoses are therefore generalizable"*[7].

Challenge and Expand Scientific Knowledge: Relate to the Literature

Subsequently, to broaden the scope of discussion according to the inverted cone analogy, the research should be compared and contrasted with other studies to add to existing knowledge[4]. Questions left unanswered by previous studies may have driven the current research; all such work which support and strengthen the authors' findings should be comprehensively cited. However, a cluttered description of all similar studies should be avoided. Earlier research is best covered in a well-articulated manner, aiming to reflect a clear set of ideas that enable logical conclusions. One technique is to construct separate paragraphs for each salient finding, covering relevant literature and concluding with a specific summary point[9]. While the key findings are discussed in this manner, care is taken to avoid discussing each and every finding[3].

Appropriate and scholarly referencing is important to validate the present research and credit the work of others[10]. While results should be discussed in the context of similar findings from prior reports, it is important not to overlook findings that are contrary to published literature[5]. Conflicting results should be mentioned with transparency, and an attempt to determine the cause(s) of contrary data, which might include differences in design, methods and patient population. If a reason for the difference is not found, this should be acknowledged clearly.

Clinical Implications and Suggestion for Further Research

The fundamental goal of all medical research is to improve patient-care. Therefore, when applicable, practical and clinical implications of the study should be included. Recommendations for practice change may be offered[5]. In the study cited earlier[6], the authors write: *"We suggest that in early CKD clinicians should understand the limitations of the creatinine-based equation, and preferably use cystatin-C based equations"*[6].

Next, the author should attempt to provide a specific agenda of research that would further our understanding on the subject[3]. While making such suggestions, it should be remembered that the hypothesis that was tested, the outcome measures, and the statistical analyses have nothing to do with the concept of requiring further testing[10].

The "Take Home Message"

The discussion should conclude with a strong, concise statement(s) that summarize the study. The content and tone of the concluding sentences should be congruent with the rest of the manuscript. The final perspective is presented, without overstating or introducing any new information. The conclusion may also provide suggestions for practice change, if relevant. In examples given above, the concluding sentences were:

- *"Therapy with tacrolimus and low-dose prednisone should be preferred to cyclophosphamide as the initial therapy for patients with steroid-resistant nephrotic syndrome, as it is effective and safe in inducing and maintaining remission of proteinuria"*[7].
- *"In conclusion, in critically ill children, withholding parenteral nutrition for 1-week while administering micronutrients intravenously was clinically superior to providing early parenteral nutrition to supplement insufficient enteral nutrition"*[8].

ELEMENTS TO AVOID WHEN WRITING THE DISCUSSION

Scientific writing needs to be objective, modest and pragmatic. The results must be supported by data, and it is important to not go overboard while interpreting the results[5]. While the overall purpose of discussion is to justify the contribution made by this study to scientific literature, the writing should not appear exaggerated or pompous[5]. Large claims should be avoided; statements such as "first time", "all the time", "wholly explains" might inflate the actual worth of the results, and do not go well with the reviewers or readers (Table 7.1) The writing should be respectful and should not attack or criticize past or contemporary researchers[5].

The discussion must focus on key messages that should be defined beforehand; tangential issues tend to dilute the major

Table 7.1: Things to avoid in discussion	
Avoid	Comment
Rewriting the results	Do not restate your results; discussion should focus on interpretation. Use bridge sentences that relate the result to interpretation.
Introducing new results	Do not present new findings in discussion.
Discussing every finding	Emphasize the key findings in a lively, crisp manner that holds attention; avoid tangential issues that distract readers.
Lengthy text	Length should be 20–25% of the overall text; avoid subtitles; references should be contemporary and focused.
Academic arrogance	Avoid claiming priority and alluding to work that is not yet completed.
Criticizing other workers	Findings that are contrary need to be handled professionally.
Over interpretation	Do not inflate importance or read more into the findings than supported by the data.
Unnecessary speculation	Emphasize the part that is a speculation, and not supported by data. Highlight need for further research.

conclusion[5]. It is often tempting to report results of unplanned analysis, deviating from the original hypothesis, when the primary results are not as expected[10]. However, no new information should be introduced in the discussion and results as obtained from the original hypothesis must be reported as such. While discussion of data derived from post-hoc analyses is acceptable and might serve as the trigger for future research, it is useful to describe such issues with clarity for the readers.

COHERENT COMMUNICATION OF KEY MESSAGES

An efficient and stimulating discussion is one that is written with the readership in mind. Most readers of medical journals, have limited time for reading, therefore read rapidly and may miss or misinterpret ideas at points that slow their reading.[12] This also implies that readers should be easily able to follow the author[s] train of thought, without having to decode or reconstruct the manuscript[12]. Clarity of expression is crucial, and imprecise words are better avoided. Authors should resist using complicated and less familiar words, avoid wordiness and delete unnecessary adverbs or adjectives. Phrases like "it is interesting to note that", "it is often the case that", "has the capability to", "is of the opinion that" and "is unable to" can be replaced by: interestingly, often, can, believes and cannot.

Brevity allows readers to capture more information in a given time[13]. Vital ideas should be expressed by short and well-constructed sentences that have more impact. Each paragraph should convey a single major point which is clearly brought forward in the first sentence, and the idea is then developed through the paragraph[12]. Continuity must be ensured between paragraphs; sentences should be ordered logically to ensure a smooth flow of ideas. Sub-sectioning interrupts this flow; therefore the discussion is best written unstructured, with transitions for moving between ideas[10]. Extensive paragraphing that interrupts flow and hinders communication of the main messages should be avoided. Clarity and conciseness often go hand-in-hand: writing that focuses directly on a point and maximizes meaning with minimum wordiness tends to be both clear and concise.

PERFECTING THE DISCUSSION

The discussion invariably requires careful editing, reconstruction and several rounds of writing attempts until real messages emerge. Two strategies which help to overcome writing flaws are described. A colleague, not involved in the manuscript, can be requested to read and suggest areas that need clarity[11]. Another strategy is to distance oneself from the first draft after writing it, regain some objectivity, re-read the relevant literature and then review the manuscript again. On returning to the manuscript, authors are bound to discover flaws missed in the first draft[13]. Additionally, authors should also consult the website of the concerned journal for additional suggestions on writing the discussion e.g., use of sub-headings or key messages in a text-box.[14]

Practice makes perfect, and scientific writing is no exception. A concise, convincing and meticulous discussion with scholarly referencing is the key to a lasting impression.

REFERENCES

1. Dewan P, Gupta P. Writing the Title, abstract and introduction: looks matter! Indian Pediatr. 2016; 53: 235–41.

2. Mukherjee A, Lodha R. Writing the results. Indian Pediatr. 2016;53: 409–15.

3. Drotar D. Editorial: How to write an effective results and discussion for the Journal of Pediatric Psychology. J Pediatr Psychol. 2009; 34: 339–43

4. Annesley TM. The discussion section: your closing argument. Clin Chem. 2010;56: 1671–4.

5. Hess DR. How to write an effective discussion. Respir Care. 2004;49: 1238–41.

6. Hoberman A, Chesney RW; RIVUR Trial Investigators. Antimicrobial prophylaxis for children with vesicoureteral reflux. N Engl J Med. 2014; 371:1072–3.

7. Gulati A, Sinha A, Gupta A, Kanitkar M, Sreenivas V, Sharma J, et al. Treatment with tacrolimus and prednisolone is preferable to intravenous cyclophospha-mide as the initial therapy for children with steroid-resistant nephrotic syndrome. Kidney Int. 2012; 82: 1130–5.

8. Fivez T, Kerklaan D, Mesotten D, Verbruggen S, Wouters PJ, Vanhorebeek I, et al. Early versus late parenteral nutrition in critically ill children. N Engl J Med. 2016;374: 1111–22.

9. Sanli O, Erdem S, Tefik T. How to write a discussion section? Turk J Urol. 2013;39 (Suppl 1): 20–4.

10. Foote M. The proof of the pudding: how to report results and write a good discussion. Chest. 2009;135: 866–8.

11. Kotsis SV, Chung KC. A guide for writing in the scientific forum. Plast Reconstr Surg. 2010;126: 1763–71.

12. Derish P, Eastwood S. A clarity clinic for surgical writing. J Surg Res. 2008;147: 50–8.

13. Morgan PP. Perfecting the manuscript: getting the right words in the right place. Can Med Assoc J. 1983;128: 769–73.

14. Indian Pediatrics. Instruction to Authors. Available from: *http://indianpediatrics.net/author1.htm*. Accessed August 21, 2016.

Ethical Issues in Biomedical Publication: A Primer for Authors

•Jaya Shankar Kaushik •Devendra Mishra

Scientific work must be conducted and reported honestly, and authors need to be aware of the ethics of scientific writing and responsible conduct. Ethical guidelines related to biomedical publications have been laid down by various scientific organizations, the primary one for medical journal authors being those by the International committee of medical journal editors (ICMJE),[1] whereas some others provide avenues to address ethical transgressions by authors.[2,3] Another good resource is the Council of science editors' white paper on promoting integrity in publication of scientific journals.[4]

The erstwhile Medical Council of India (MCI), while making it mandatory to have research publications for appointment as medical teacher, also provided guidelines about the paper-category and the author sequence, which would qualify for their criterion.[5] This has brought in a certain pressure to publish more, both among medical teachers, and those aiming to be one. Under this pressure to perform, fundamental ethical principles of publishing are sometimes circumvented. A survey of biomedical researchers reported that only a minority had substantial knowledge about ethical issues in publication.[6] Another survey reported scientific misconduct to be commonly observed among Indian researchers, major ones being gift-authorship (65%), data-falsification (53%), and plagiarism (53%).[7]

We, herein provide a primer for authors on common ethical issues in biomedical publication, with the intention of addressing the lack of awareness regarding such issues among young researchers/authors.

COMMON ETHICAL ISSUES

Fabrication or Falsification of Data

Falsification is manipulating data or experimental procedures to produce the desired outcome or to avoid a complicating or inexplicable result. In contrast, fabrication is defined as recording or presenting (in any format) fictitious data. In simple terms, if authors report data that was never collected, it is a fabrication; if they manipulate or change the collected information, it is called falsification. A frequently encountered scenario is authors modifying the reported methodology based on the reviewers' comments, after rejection by that journal. Subsequently, it is submitted to a different journal (falsification). Any change in the reporting of study design, study instruments, methodology or results of the research, once it has been done in some other way, amounts to falsification.

If authors did not collect data at all, or data from a limited number of subjects and reported it for a much higher number, it is called fabrication of data. Similarly, if authors report the use of a technology or a tool to collect the data, which they did not do, it amounts to fabrication.

There have been a large number of reports of such "data fraud" in clinical trials.[8] As a rule, editors of the journal consider the authors to be honestly reporting their findings and morally responsible for what they are submitting for publication. At the journal's level, it is challenging to detect falsification or fabrication of data unless some whistleblower informs the journal. Occasionally, expert reviewers of various journals overlap, and they may red-flag such misconduct. Identification of such misconduct by a researcher may invite punitive action from the parent institute, ban on receiving funds from funding agencies, or even legal action.[9]

Plagiarism

The term plagiarism is derived from the Latin word "plagiarius" meaning kidnapper. Plagiarism is defined as "appropriation of another person's ideas, processes, results, or words without giving appropriate credit."[10] It is primarily stealing someone else's work and claiming it to be yours. It could be a word to word copy of the text, or it could be paraphrasing the sentences. Most authorities agree that copying more than six consecutive words from another publication would qualify as plagiarism.[11] Using someone else's idea without appropriate citation also amounts to plagiarism, though it is somewhat difficult to prove. Even if you use text from your own previous publication without citation, it counts as plagiarism (self-plagiarism). Plagiarism among novice authors is often due to ignorance rather than an actual intent to copy—for authors whose primary language is not English often find it easier to use published sentences/phrases to express their ideas instead of framing their own sentences. Few suggestions to prevent unintentional plagiarism are provided in Box 8.1.

Plagiarism is a form of academic dishonesty and needs to be avoided by authors and strictly acted against by editors. There are many plagiarism detection softwares that can detect the extent of duplication between a manuscript and pre-existing published material. Many reputed journals routinely screen all submitted papers for plagiarism, whereas others do it on a case-to-case basis.

Box 8.1 **Writing practices to guard against plagiarism concerns**

- Use a quotation mark when you must use significant portions of already published material (e.g. standard definitions, standard methods or hypothesis/ideas of others).
- Cite appropriate references for any data/idea adapted from another source.
- Avoid using cross-references (citing an article based on somebody else's citation); read the source article in full.
- If an idea generated from a talk that you heard in a conference or from personal communication with any other researcher, it is advisable to mention the same in the manuscript.
- Students in their early career life often tend to copy-paste from online resources without any wrong intent—the senior author must take appropriate measures to prevent the same.
- Copyrighted material (e.g. tables, figures) cannot be reproduced without permission. Citing a reference alone cannot justify the infringement of copyrighted material.
- If in doubt, screen your manuscript for plagiarism using available softwares before submission—many of these are available online and are free to use.
- Reading the copyright form of the journal before signing provides novice authors a good insight into what all they are agreeing to while submitting a manuscript.

Authorship Issues

Authorship conflicts are common among researchers.[7] The ICMJE[1] has very succinctly provided the criteria for authorship for a biomedical publication, "should have made substantial contributions to conception and design, or acquisition of data, or analysis and interpretation of data; drafting the article or revising it critically for important intellectual content, final approval of the version to be published; and willingness to take public responsibility for all aspects of the study." It has been observed that almost half of the manuscripts submitted to individual journals do not qualify as per these criteria.[12] There are two aspects to this problem *viz.*, granting authorship to people who do not qualify authorship criteria (gift authorship),[13] and authors who have contributed significantly in the research and fulfil the authorship criteria, but are missing from the list of authors (ghost authors).

Few sound principles of authorship, so as to avoid conflicts later, are provided in Box 8.2. It is the amount of work done by the individual researchers, and their fulfilling authorship criteria, which helps in deciding how many authors a publication will have. As a general guide, a maximum of 3-4 authors from one department

---

Box 8.2   **Healthy practices for authorship of research papers**

- Decide the sequence of authors *a priori* at the time of the research plan/ protocol, based on their level of contribution, according to a pre-decided scheme.
- The research student who collects the data, analyzes it, and writes the first draft of manuscript should usually be considered for first authorship; even though, some others may contend that it is the concept or the idea which is more important.
- The senior-most author, who guides and supervises the research work and is responsible for the authenticity of data, should preferably be the corresponding author.
- Administrative heads that otherwise had no role in the research should be acknowledged for their support, and should not be gifted authorship.
- If you agree to be a co-author to any research paper, it is your responsibility to ensure you fulfill authorship criteria and agree with all material in the manuscript.
- Do not agree to accept gift authorship, or sign authorship forms when you have not contributed to a research study.
- Develop a culture of ethical publication in your institute.

and one author from each collaborative department are good enough for papers reporting single-institution research studies. However, there is no real limit on the number of authors a paper may have.

Authorship sequence is also a common sore point among authors, especially so after the MCI guidelines and guidelines of various universities/appointing authorities, wherein some authors are given more credit than others for the same paper (e.g., first author, corresponding author, first two authors). By convention, the first author is expected to have provided the maximum contribution to the paper, and consequently first authorship gets more credit. Although there has been much discussion on the issue, there are no clear guidelines for setting the sequence of authors, but deciding it before the start of paper-writing or even before the start of the research is the ideal way to avoid disputes later-on.[5,14] It is our considered opinion that all authors are equally important, irrespective of their position in the byline, and interview boards or evaluation bodies should provide equal credit to all authors.

Conflict of Interest

A researcher might have multiple interests apart from publishing a research paper. In an ideal world, we expect that any other financial or technical interests of the researcher will have no bearing on the direction of results of the research. If this is not the case, we fear that one interest might have a conflict with the primary interest in question i.e., research in this case. In technical terms, Conflict of interest (COI) is defined as *"a set of conditions in which professional judgment concerning a primary interest (such as the validity of research) tends to be unduly influenced by a secondary interest (such as financial gain).[15] For example, COI is apparent when the investigator in a drug trial is either closely related to the manufacturer or has shares in the company or has previously received funds from the company for some scientific activity."*[16]

Any possible conflict of interest must be disclosed by the authors at the time of submitting the manuscript. The vital point here is that while declaring, it does not matter whether any actual COI occurred—what is important is declaring even any potential for a conflict. Anything that is not declared at the outset and discovered after the article gets published results in embarrassment, and may also lead to questioning of the integrity of the researcher.

Salami and Duplicate Publication

Traditionally, salami is a sausage that is sliced into multiple pieces for eating. Similarly, when a single research data collected over the same period is sliced into various papers, it is called salami publication. For example, in a study on children with neurocysticercosis, if radiological findings are published in a radiology journal and clinical features in a separate journal without each paper citing the other, this constitutes salami publication. However, clear-cut definitions for this transgression and avenues to address the same are not as forthcoming, as for other publication misconduct. Voluminous cross-sectional community-based data or prospectively collected data over a long period of time may provide opportunity for publication of multiple papers addressing inter-related issues, and authors need to take care of always citing the main study and informing the editors about the related previous publications.

Submitting/publishing identical or substantially overlapping material at another place by the same author (with/without acknowledgment to the previous publication) is called duplicate submission/publication. Duplicate reports are often retracted by the journal; in addition, the researcher may be banned from submitting to that or a group of journals, other journals of the specialty may be intimated, and even the author's institution may be informed of such academic misconduct.

AVOIDING ETHICAL TRANSGRESSIONS

Publication ethics are a set of ethical/moral behaviors that are expected from authors, reviewers, editors and publishers, and from academic medical institutions. Accepted publication standards have been laid down by many organizations, the premier one being from the ICMJE.[1] Adequate care needs to be taken to avoid unethical conduct, not to think of them as a mere formality while submitting the manuscript. We feel that the fundamental ethical principle in research is truthfulness and honesty. Authors in the early parts of their careers should avoid presumptions and when in doubt, they should ask a senior colleague or an expert. Writing clearly to the editor of the journal you are submitting is another fail-safe method to address any ethical dilemma related to publication. Being a co-author to an unethical work is an equally heinous crime in medical publishing. So do not run behind numbers, credits, and promotions, at the cost of basic ethical principles. One dark spot of unethical publication and public shaming is enough to ruin the entire career of a researcher.

CONCLUSION

It is essential to adhere to ethics in biomedical publication, especially in this era where publication or patents are the only measures used for academic performance. All authors, reviewers, editors and medical publishers are bound by specific moral responsibilities. Honesty and truthfulness should prevail in medical science. All authors should be aware of these fundamental ethical principles before submitting manuscripts for publication in biomedical journals.

REFERENCES

1. International Committee of Medical Journal Editors. Recommendations for the Conduct, Reporting, Editing and Publication of Scholarly Work in Medical Journals. Available from: *http://www.ICMJE.org*. Accessed on March 8, 2019.
2. Ferris LE, Fletcher RH. Conflict of interest in peer-reviewed medical journals: the World Association of Medical Editors (WAME) position on a challenging problem. Natl Med J India. 2010;23: 65–8.
3. Wager E. The Committee on Publication Ethics (COPE): Objectives and achievements 1997-2012. Presse Medicale Paris Fr. 2012;41: 861–6.
4. Editorial Policy Committee, Council of Science Editors. CSE's White Paper on Promoting Integrity in Scientific Journal Publications. Wheat Ridge, CO: 2018. Available from: *www.councilscienceeditors.org/resource-library/editorial-policies/white-paper-on-publicationethics/*. Accessed on 18 March, 2019.
5. Aggarwal R, Gogtay N, Kumar R, Sahni P, for Indian Association of Medical Journal Editors. The revised guidelines of the Medical Council of India for the academic promotions: Need for a rethink. Indian Pediatr. 2016;53: 23–6.
6. Schroter S, Roberts J, Loder E, Penzien DB, Mahadeo S, Houle TT. Biomedical authors' awareness of publication ethics: an international survey. BMJ Open. 2018;8:e021282.
7. Dhingra D, Mishra D. Publication misconduct among medical professionals in India. Indian J Med Ethics. 2014;11:104–107
8. George SL, Buyse M. Data fraud in clinical trials. Clin Investig. 2015;5: 161–73.

9. Resnik DB. Data Fabrication and Falsification and Empiricist Philosophy of Science. Sci Eng Ethics. 2014;20: 423–31.

10. U.S. Department of Health and Human Services (HHS). Public Health Service policies on research misconduct. Final rule. Fed Regist. 2005;70:28 369–400.

11. Masic I. Plagiarism in scientific publishing. Acta Inform Medica. 2012;20: 208–13.

12. Šupak-Smolčié V, Mlinarié A, Antončié D, Horvat M, Omazié J, Šimundié A-M. ICMJE authorship criteria are not met in a substantial proportion of manuscripts submitted to Biochemia Medica. Biochem Medica. 2015; 25: 324–34.

13. Juyal D, Thawani V, Thaledi S, Prakash A. The fruits of authorship. Educ Health. 2014;27:217.

14. Bhattacharya S. Authorship issue explained. Indian J Plast Surg. 2010; 43: 233–4.

15. Thompson DF. Understanding financial conflicts of interest. N Engl J Med. 1993;329: 573–6.

16. Ghooi RB. Conflict of interest in clinical research. Perspect Clin Res. 2015;6:10.

9

Accurate References Add to the Credibility

• Anup Mohta • Medha Mohta

While planning a clinical research and subsequently writing the manuscript, scientific papers on the related subject are often retrieved and articles in their reference lists are sought to search more relevant literature. Occasionally it is very difficult to trace some of these articles and almost impossible to access certain internet citations due to various errors in citations by the authors. Many studies have been conducted to assess the extent of this problem, and a variable frequency of errors in writing the references has been reported[1-5]. Importance of accuracy of citations and references in research publications cannot be overstressed, given their role in leading the reader to the relevant literature.

Every researcher who wishes to carry out and publish authentic material needs to (a) look for pre-existing literature on subject; (b) review and analyze the studies in the context of the proposed research; (c) evaluate, compare and/or contrast existing information against the findings of his own results and explain the differences; (d) prepare a suitable manuscript for the target audience; and (e) acknowledge all sources of information, which have been utilized in the preceding steps[6].

WHAT ARE REFERENCES?

The term 'References' is used for the list of sources that have been accessed by the author and cited in the manuscript. This could include journal articles, books or book chapters, monographs, internet sites and other sources. On the other hand, the term 'Bibliography' is the term commonly used for a list of sources of information (e.g. books, articles, and internet sources) used to

First published as Mohta A, Mohta M. Accurate References Add to the Credibility. Indian Pediatr. 2016; 53:1003–06.

prepare the manuscript, and it includes all the sources accessed even if they have not been referred to in the manuscript.

Reference list is placed after the text of the manuscript starting on a fresh page, whereas bibliography is not provided with majority of journal articles.

WHY REFERENCES?

The author needs to introduce the research question in view of the available literature on the subject, for which pre-existing literature is the cornerstone of the argument. Moreover, methodology used to carry out the research would include procedures and other tools described previously by other authors, and would need to be cited. Finally, results of the research need to be evaluated, interpreted, and discussed in relation to the previously conducted studies. These steps in preparation of the manuscript need literature search, which when cited, become the part of the reference-list of the manuscript.

Apart from giving due credit to previous researchers, the reference-list allows the readers to verify the methodology from the original source. Moreover, the reader can evaluate the results in the context of the cited literature and verify whether the authors have interpreted and cited earlier studies correctly.

A complete reference list also makes it convenient for the editorial board of the target journal to select and invite reviewers for the manuscript as they refer to the reference list for scientists with work in the relevant field. It may also help the editors and reviewers to check for possible plagiarism from the cited sources of information.

WHERE AND HOW MANY?

As a rule of thumb, literature should be cited wherever the author has used the idea, methodology, data, figures or diagrams from a previously published source. This could be in the Introduction, Methodology, Tables, Figures, and Discussion of the manuscript. It is essential to include key references relevant to the study. Ignoring to cite consulted literature may amount to plagiarism. Abstract and Results sections usually do not have any references.

Number of references to be cited should be in accordance with the Author Guidelines issued by the target journal. Many journals do not prescribe any limit of references for the research or review papers but do have limits for certain sections like Case reports, Correspondence and Images. Due to paucity of print space available, it is advisable to use only the most relevant and recent references. Very old references should be avoided as they may not be available to reviewers and readers to access and read. It is not necessary to reach the maximum limit if authors can convey the message with lesser number of references.

When to organize the reference list: It has been suggested that the authors should organize the reference list when the manuscript is being organized[7]. This allows the author to: identify which previously published article can be best cited in the manuscript; rationalize the selection of the adequate references and avoid bias towards one kind of viewpoint; and limit the number of references to meet the requirements of the target journal.

Styles of References

Most medical literature uses two formats of references, i.e. Harvard style and the Vancouver style. Harvard style is used by many journals of basic sciences and details on Harvard Style can be accessed elsewhere[8]. Vancouver style[9] is followed by most medical journals, including *Indian Pediatrics*. Basic differences in two styles are summarized in Box 9.1.

Vancouver Style

There are two components of citation: In-text citation and the Reference list.

In-text citation

- All references are cited and numbered sequentially as they are referred to in the text. If the reference is cited in Table or figure, the citation number should be given consecutive to that in the preceding text.
- If a source has been cited and identified by a number, same should be used consecutively throughout the manuscript if the same source is cited again.
- Numbers may be placed in parentheses or superscript in accordance with the Journal policy.

Box 9.1	Differences between the Vancouver style and the Harvard style

Vancouver style
- Sequential in-text citation; identified by an Arabic number (in parentheses or superscript).
- Reference list: Numbered sequential list in order of citation in text.

Harvard style
- In-text citation by authors name and year of publication. Page numbers may be included for being more specific.
- Reference list: Alphabetical order without numbering

- Identifying numbers are placed after punctuation marks like full stops or commas, and before colons and semi-colons[9]. However variations can be there in different journals.
- If two or more sources are cited simultaneously, they should be identified in chronological order according to their date of publication by numbers separated by a comma. They should be mentioned in chronological order with date of publications
- If several consecutively numbered sources are cited, hyphen should be used instead of comma; e.g. [3–6] instead of [3,4,5,6] .

Reference list

Aim of the reference list is to provide complete information for access of the cited source.

- The list is prepared in sequentially numbered order in which the source has been cited in the text. Citation number in the text and the list should match.
- Last name of the authors are written first followed by the initials. If there are six or fewer authors, names of all the authors are included. In case there are more than six authors, names of first six authors are listed followed by et al.
- Abbreviated names of the most medical journals are available in Pubmed's Journal Database available at *http://www.ncbi. nlm.nih.gov/nlmcatalog/journals.* In case of a non-medical Journal or that which is not available on PubMed, other data bases or journal website can be looked.
- First letter of each author's last name and initials should be in capital letters without any intervening full stops. Capital letters should also be used for the first letter of the publication title, and all first letters of a place name and publisher.

Method of citing common sources of information in the reference list has been shown in Box 9.2. detailed information on citing various sources of information is available on National library of medicine[10] and is recommended by international committee of medical journal editors[11]. There are minor variations from the basic format as some journals expect full names of the journals, name of the journal in italics, volume number in bold letters etc. Therefore, it is necessary for authors to check the author guidelines and give due attention while citing in the text and preparing the reference list[12]. Editors appreciate these as it suggests that the author has read the Instructions to authors diligently.

Box 9.2 | **Citing references as per Vancouver style**

Single author
Mohta A. Common conditions in pediatric surgery. Indian J Pediatr. 2014;81:684-9.

Multiple authors (less than six authors)
Falagas ME, Korbila IP, Giannopoulou KP, Kondilis BK, Peppas G. Informed consent: how much and what do patients understand. Am J Surg. 2009;198:420-35.

Multiple authors (more than six authors)
Dillon P, Hammermeister K, Morrato E, Kempe A, Oldham K, Moss L, et al. Developing a NSQIP module to measure outcome in children's surgical care: opportunity and challenge. Semin Pediatr Surg. 2008;17:131-40.

Book
Townsend CM, Beauchamp RD, Evers BM, Mattox KL. editors. Sabiston textbook of surgery-the biological basis of modern surgical practice. 17th ed. Vol 1. New Delhi: Saunders; 2004.

Chapter in a book
Argani P, Beckwith JB. Renal neoplasms of childhood. In: Mills SE (Editor). Sternberg's diagnostic surgical pathology. 4th ed. Philadelphia: Lippincott Williams and Wilkins; 2004. p. 2029-30.

Internet reference
World health organization. The world health report. Available from: URL: http://who.int/whr/en/index.html. Accessed september 19, 2009.

Author Names

Citing the name of the authors correctly in text and reference list is essential. Names of the authors should be taken from the original publications. Only the last name of the author is cited in the text without initials. If the cited paper has two authors, both names may be cited e.g. Gupta and Chawla; if more than two authors are listed, last name of the first author followed by 'et al.' is adequate.

Abbreviation of the names of the authors has been discussed widely. Many errors appear in the names as we are not aware of the pattern of writing the names. It is a common practice to write the first name last in chinese authors[13]. Similarly, in authors from southern states of India, first name is the last word preceded by some initials which may refer to the family name or that of the village. It may be difficult to abbreviate the names. Some names may include special characters or letters which again cause problems. Long names are difficult to abbreviate. If the full name of the author is Lakshmi Kant Radha Reddy, it can be abbreviated variously as Reddy LK, Reddy LR, Radhareddy LK as per the perception of the author preparing the manuscript. Similarly, spellings of Chaudhary can be also be written as Chaudhury, Chowdhury, Choudhury etc; all of which shall not be available in a single search in a database. If some scientist needs to retrieve all the scientific work done by these authors, it will be impossible due to incorrect abbreviation of name or misspelt last name[14]. Advice may be taken from an author from same cultural background to correctly abbreviate the name. Some journals give the citation to be used with abbreviated names to avoid errors.

Sometimes the difference in the names of the authors in various sources appears due to carelessness by the authors while submitting their manuscripts. It has been seen that the name of same author has been differently mentioned even in the credits of two papers in the same issue of the journal[15]. Therefore, it is incumbent on the authors to check their full names correctly with desired abbreviated names so as to prevent any errors.

COMMON PITFALLS

Although many authors take due care in preparation of the reference list and exercise due diligence in verifying the same, sometimes errors crop up in the citation in text as well as in the reference list.

Errors in Citation

1. Citing a reference as a cross reference and not reading the original article
2. Citing the same citation more than once in a manuscript leading to duplication
3. Not identifying the in-text citation as per the format of the target journal
4. Citing the name of the authors along with the initials
5. Pooling multiple citations at the end of the sentence instead of insertion along with the referred fact
6. Citation in text without appropriate entry in the reference list
7. Citing a reference after reading only the abstract
8. Personal communications as reference
9. Inclusion of references in abstract and results section

Errors in Reference List

1. Not in the desired format of the journal
2. Copied from another article without direct verification from original article

3. Copied from scientific data bases without reading the original article[16]
4. Errors in abbreviation of names of authors, journal, page numbers, volume number etc.
5. Error in number of authors listed
6. Inclusion of non-essential items like issue number, or date and month of publication.
7. Submission of a previously rejected article to new journal without modifying the citation and reference list style in accordance with the new Journal.
8. Citation of print version while the internet version has been accessed
9. Inclusion of retracted references

RESOURCES FOR IMPROVING CITATION AND REFERENCES

Reading the original article for citation and inclusion in reference list as per the journal guidelines is the most important requirement for accuracy. However, some assistance is available for improving the same.

Many reference and citation management softwares are available for scholars for recording and utilizing citations or references. This can be used to reproduce the reference lists at more than one occasion. Some of these softwares are EndNote, and reference manager by thomson reuters; papers and readcube by labvita; mendeley by elsevier and zotero[7].

Many professional manuscript submission systems use software to validate the uploaded references. If a reference cannot be validated, authors need to verify the source of the same.

CONCLUSION

It is essential to cite the original literature reviewed and used for preparation of the manuscript, and provide a correct reference list. It acknowledges the previous researchers, and also gives credence to your work. Due diligence exercised in citation and preparation of the reference list makes the editors and reviewers confident about the research and the manuscript. As the paper becomes more credible, the readers are more convinced about the study and shall be able to cite it with confidence. Although apparently taxing, efforts made for citing and listing appropriate references may increase the acceptability of the manuscript. The recent availability of reference management software (e.g., Menedeley, Zotero) has made this work easier to accomplish, though selecting the appropriate reference, and reading thorugh it, is still the author's responsibility.

REFERENCES

1. Asano M, Mikawa K, Nishina K, Maekawa N, Obara H. The accuracy of references in Anaesthesia. Anaesthesia. 1995;50: 1080–82.
2. Mohta A, Mohta M. Accuracy of references in Indian Journal of Surgery. Indian J Surg. 2003;65: 156–8.
3. Aronsky D, Ransom J, Robinson K. Accuracy of references in five biomedical informatics journals. J Am Med Inform Assoc. 2005;12: 225–8.
4. Gupta P, Yadav M, Mohta A, Choudhury P. References in Indian Pediatrics: Authors need to be accurate. Indian Pediatr. 2005;42: 140–5.
5. Wager E, Middleton P. Technical editing of research reports in biomedical journals. Cochrane Database Syst Rev. 2008;4:MR000002.
6. International Baccalaureate. Effective citing and referencing. 2014. Available from: *http: //www.ibo.org/globalassets/digital-tookit/brochures/ effective-citing-and-referencing-en.pdf.* Accessed May15, 2016.
7. Annesley TM. Giving credit: Citations and references. Clin Chem. 2011; 57: 14–7.
8. University of Sydney. Your guide to Harvard Style referencing. Available from: *https://www.library.usyd. edu.au/subjects/downloads/citation/ Harvard_Complete.* pdf. Accessed April 25, 2016.
9. Monash University Library. Vancouver Referencing and Citing Style. 2015. Available from: *http://guides.lib. monash.edu/citing-referencing/vancouver.* Accessed May 16, 2016.
10. International Council of Medical Journals Editors. Recommendations for the Conduct, Reporting, Editing, and Publication of Scholarly Work in Medical Journals. Updated December 2015. Available from: *http:// www.icmje.org/icmje-recommendations.pd*f. Accessed May 16, 2016.
11. Patrias K. Citing Medicine: The NLM Style Guide for Authors, Editors, and Publishers. 2nd ed. Wendling DL, technical editor. Bethesda (MD): National Library of Medicine (US); 2007. Available from: *http://www.nlm. nih.gov/citingmedicine.* Accessed May 15, 2016.
12. Peh WCG, Ng KH. Preparing the references. Singapore Med J. 2009;50: 659–62.
13. Zhaoping L. For your name's sake. Curr Biol. 2010;20:R341.
14. Todd PA, Ladle RJ. Hidden dangers of a 'citation culture'. Ethics Sci Environ Polit. 2008;8: 13–6.
15. Mohta A. Responsible authorship. Indian J Urol. 2011;27: 288–9.
16. Franceschini F, Maisano D, Mastogiacomo L. The museum of errors/ horrors in Scopus. J Informetr. 2016;10: 174–82.

Sharing Clinical Experience with the Scientific Community: How to Write a Case Report?

• Sumaira Khalil • Devendra Mishra

"Always note the unusual…. Publish it; Place it on personal record as a short note. Such communications are always of value."[1]

This quote by Sir William Osler, the father of Modern Medicine, is a testimony to the continual teaching value of case reports. A case report is a description of a single case or cases with unique features experienced by physicians, which are then shared with the scientific community.[2] These are the first line of evidence in healthcare system, and are important in communicating something new learnt from clinical practice.[3]

WHY WRITE A CASE REPORT?

We all know that case reports are ranked quite low in the hierarchy of evidence-based medicine[4], but still retain value in scientific publishing as an important mode of providing new knowledge to the field of medicine. They provide a great opportunity for young researchers to develop their writing skills, perform literature search, experience the peer-review process, and start their scientific writing careers[5,6]. They are instrumental in formulating novel ideas that trigger clinical trials for future research. Albrecht, et al.[7] analyzed all the case reports and case series published in the Lancet from January 1996 to June 1997, and found 23.3% (24/103) to be followed by randomized controlled trials on the same topic. Case reports are the only medium for reporting unique cases, rare associations, atypical presentations, and unexpected outcomes. Reporting adverse drug reactions by means of case reports are fundamental part of pharmacovigilance[8]. Both the first heart transplant by Christian Bernard in 1967, and the first ever report of people with

First published as Khalil S, Mishra D. Sharing Clinical experience with the scientific community: How to write a case report? Indian Pediatr. 2016; 53:513–16.

Box 10.1 Common clinical experiences which may be shared by case reports

- To report a rare or unknown disorder
- To describe an atypical etiology or presentation
- To discuss a challenging differential diagnosis
- Rare manifestations/complications of a known disease
- Rare associations with sound justification
- New insights into pathogenesis of a disease
- To report unusual drug-interactions
- To describe new/rare side-effects of drugs
- To prompt or disconfirm a hypothesis
- To report any novel diagnostic procedure
- To report a new treatment modality
- To report an unexpected outcome
- Common topics with high clinical relevance

Acquired Immunodeficiency Syndrome (not recognized as such yet) in 1981 were published as case reports.[9,10] The various situations where one may think of publishing a case report are illustrated in Box 10.1.

STRUCTURE OF A CASE REPORT

A case report should be a structured, brief and focused document highlighting a clear learning-point. Case reports do not follow the IMRAD (Introduction/Methodology/Results/Discussion) structure that is followed in research articles, but still follow the underlying schema of Why (you are reporting the case?), How (it was done?), What (was found?), and What (it all means?). There is minor variability amongst journals, but mostly the following sections are present in a case report:

Title: The title of the case report is obviously one of the first thing seen by the reader, reviewer or editor. The title should give them an idea regarding what is being reported. The title must not describe all the findings, but create a sense of curiosity amongst the readers. Therefore, it should be simple, specific, concise, catchy and eloquent.[11,12] Some of the journals, and the consensus-based clinical guidelines case-reporting (CARE)[13], suggest that the title contains the words 'case report', e.g., 'Headache and transient visual loss as the only presenting symptoms of vertebral artery dissection: a case report[14].'

Usually, it is preferable to write the title after writing the whole manuscript, as it is at that time that one has a better overall impression of the key message to be conveyed.

Abstract: It is a brief and short description (50–100 words) emphasizing the prudent points of the case, and anything new added to the existing knowledge. Abstracts for case reports are generally unstructured, and some journals might not even ask for an abstract. However, a 4-point structured abstract is required by *Indian Pediatrics,* developed after discussions among a variety of stakeholders.[15] The CARE guidelines also have recommended a structured abstract.

Introduction: The introduction should be short, and describe the background regarding the case to be reported, and what is already known. A justification regarding why we need to report this has to be given with what has already been reported in the past and what new will it add to the existing knowledge. Some journals do not even ask for an introduction and the manuscript starts with the description of the case.[16]

Case-description: The description of the case must report all events in a chronological order. This section describes the history, clinical examination findings, demographic data and all investigation reports that support the diagnosis, and exclude the other differential diagnosis. The treatment and follow-up must be described in detail, and all important negative findings must also be included. The case-description must be sufficiently detailed in order to give an opportunity to the readers to make their independent clinical impression and differential diagnoses. This may be supported with figures like clinical photographs, radiological images, and photographs of pathological specimens/slides. All information conveying the case-detail must be included, while avoiding superfluous details that may break the flow.[12,16] Take care not to detail all the investigations carried out; only the relevant ones need to be detailed with actual values (preferably with normal range, if it is an uncommon or specialized investigations and majority of readers are unlikely to be aware of the normal values) and units as per the journal's requirements. Rest of the investigations carried out can be clubbed together and stated to be 'within normal limits' or by some such statement. Some journals may even allow you to add a table to detail all the investigations, if the work-up was too exhaustive, or the patient was followed-up for a long period; refer to past issues or the authors' instructions of the concerned journal for guidance.

It is preferable to provide the relevant imaging/micrograph pictures with the submission; these may or may not be included in the final publication but allow the reviewer to make an informed decision about your work-up and diagnosis. Always take a written, informed consent from the patient/parents/legal representative for the publication of the case report, including photographs and clinical details (after showing them to the patient/parents). This precludes problems later, when the patient is non-contactable after discharge, and the journal asks for a permission letter.

Discussion: This is a crucial part of the manuscript, and justifies why the case is worth reporting. This section should start with a summary of the salient features of the case, followed by comparison with similar cases reported in the past and reasons why the presented case is different. Do not claim 'first such case' as your literature search strategy may not be systematic or comprehensive, and is likely to miss similar cases reported in literature. The discussion evaluates the case for its novelty, uniqueness, variability and appropriateness, comparing with literature published in the past to derive any new knowledge and applicability in clinical practice.[2] The justification for the present diagnosis/intervention must be discussed, followed by some implications of the case on clinical practice. As case reports have a low level of evidence, we must not make overambitious generalized recommendations; rather, limit ourselves to enlist the learning points that add to existing knowledge, and make some appropriate and specific suggestions depending on the quantity of literature available[17].

References: The number of references permitted for a case report is very limited; this varies amongst journals—*Indian Pediatrics* permits a maximum of 10 references. You do not need to cite all previous such cases published, but the more relevant and preferably the recent ones.

AUTHORSHIP ISSUES

Most editors will agree that authorship issues are most commonly encountered with case reports. Both gift-authorship and exclusion of deserving authors (ghost authors) is common. This has led quite a few journals (e.g., *Indian Pediatrics, BMJ case reports*) to limit the number of authors permitted for case reports.

These authorship issues quite frequently arise as the authorship criteria provided by various groups like the ICMJE are not easily applicable to case reports. Moreover, as it is not a pre-planned 'study', very frequently the requirement 'substantial contributions to the conception or design of the work' is missing, and it is also difficult to differentiate 'acquisition, analysis, or interpretation of data' from routine clinical care. Though, the remaining three criteria remain valid. Thus, it has been suggested that one authorship criteria could be "all authors must have made an individual contribution to the writing of the article and not just been involved with the patient's care"[18].

Another needling issue is the authorship opportunity to members of investigative or supportive departments e.g., Radiodiagnosis, Pathology, Pediatric surgery. As the admitting department/unit has the access to the patient data and follow-up, supportive departments may frequently miss on the opportunity to report the case; even though it may have been their contribution that led to the diagnosis (imaging or biopsy) or improvement (surgery). There is, thus, a need for individual institutes to develop guidelines or standard operating procedures for reporting of cases and the ownership of the data. A time-limit may be set for the admitting department/unit to prepare and submit the case report, after information to all departments involved in the care of the patient. An oversight group from these departments may discuss and decide on the authors.

Till recently, there was a wide variability in guidelines for writing case reports and instructions to authors of the specific journals had to be followed. Consensus-based clinical guidelines case-reporting (CARE) have recently been proposed by Gagnier, et al.[13] The checklist provided[19] is used by many journals and reviewers while reviewing case reports. These guidelines bring about completeness in writing the case, thereby increasing their chances of acceptance.

LIMITATIONS OF CASE REPORTS

Case reports are known to have some inherent limitations. Findings are specific to that particular case and cannot be generalized, and make limited contribution to the scientific knowledge-base. Moreover, as they are lower in the level of evidence, they are cited infrequently, very often leading to a detrimental effect on the journal's impact factor. Some journals have limitation of space and have had incidents of authorship abuse in the past, which has lead to removal of the case report section from their journal[20, 21]. Many of the interview boards, and even the medical council of India[22],

Box 10.2	Common reasons for rejection of case reports

- Too common condition
- Too rare condition, that readers are unlikely to encounter (usually indicates a mismatch with the readership)
- Too obvious diagnosis
- New gene mutation but with no clinical relevance
- Diagnosis not robust/required investigations not done
- All differential diagnoses not ruled out
- Unethical investigation or treatment
- No teaching point/value
- Unclear message or wrong message

do not count publication of case reports as a 'research publication'. It is becoming more and more difficult for authors to get their case reports accepted in mainstream journals. Current acceptance rate of case reports by Indian Pediatrics is below 5%. Some common reasons for rejection of case reports are detailed in Box 10.2, which may be helpful to the beginner. However, the concept of journals dedicated only to publication of case reports is a positive change.

CONCLUSIONS

Case reports are brief excerpts where clinicians describe their experience of a particular case. Despite inherent limitations and limited educational value, they still remain an important tool for sharing scientific knowledge, and an easy avenue for polishing the writing-skills of the beginner. The persisting interest of readers in case and the arrival of many journals primarily dedicated to publication of case reports, will ensure that this important link in scientific evidence does not become extinct.

REFERENCES

1. Osler W: The Quotable Osler. Philadelphia: American College of Physicians; 2003.
2. Cohen H. How to write a patient case report. Am J Health Syst Pharm. 2006;63: 1888–92.
3. Rison RA. A guide to writing case reports for the Journal of Medical Case Reports and BioMed Central Research Notes. J Med Case Rep. 2013;7:239.
4. Papanas N, Lazarides MK. Writing a case report: polishing a gem? Int Angiol. 2000; 27:344-9.
5. Florek AG, Dellavalle RP. Case Reports in medical education: a platform for training medical students, residents, and fellows in scientific writing and critical thinking. J Med Case Rep. 2016;10:86.

6. Nissen T, Wynn R. The clinical case report: a review of its merits and limitations. BMC Res Notes. 2014; 7:264.

7. Albrecht J, Meves A, Bigby M. Case reports and case series from Lancet had significant impact on medical literature. J Clin Epidemiol. 2005; 58: 1227–32.

8. Bavdekar SB, Save S. Writing case reports: contributing to practice and research. J Assoc Phys India. 2015; 63:44-8.

9. Kantrowitz A, Haller JD, Joos H, Cerruti MM, Carstensen HE. Transplantation of the heart in an infant and an adult. Am J Cardiol.1968;22: 782–90.

10. Centers for Disease Control: Pneumocystis pneumonia – Los Angeles. MorbMort Wkly Rep.1981;30: 1–3.

11. Dewan P, Gupta P. Writing the title, abstract and introduction: Looks matter! Indian Pediatr. 2016;53: 235–41.

12. Wang YX. Advance modern medicine with clinical case reports. Quant Imaging Med Surg.2014;4: 439–43.

13. Gagnier JJ, Kienle G, Altman DG, Moher D, Sox H, Riley D and the CARE Group. The CARE guidelines: consensus-based clinical case reporting guideline development. Glob Adv Health Med. 2013;2: 38–43.

14. Yvon C, Adams A, McLauchlan D, Ramsden C. Headache and transient visual loss as the only presenting symptoms of vertebral artery dissection: a case report. J Med Case Rep. 2016;10:105.

15. Mishra D, Gupta P, Dhingra B, Dewan P. Proposal for a structured abstract for case reports: an analytical study. Poster Session Abstracts, Peer Review Congress, Chicago 2013.p.70. Available from: *http://www.peerreview congress.org/abstracts_2013. html#4*. Accessed April 28, 2016.

16. Peh WCG, Ng KH. Writing a case report. Singapore Med J. 2010; 51:10-3.

17. Alwi I. Tips and tricks to make case report. Acta Med Indonesia. 2007;39: 96–8.

18. BMJ Case Reports. Instructions for Authors. Available from: http:// casereports.bmj.com/site/about/guide lines. xhtml#author. Accessed May 02, 2016.

19. Equator Network. The CARE Guidelines: Consensus-based Clinical Case Reporting Guideline Development. Available from: http://www.equator-network.org/report ing-guidelines/care/. Accessed May 02, 2016.

20. Martyn C. Case reports, case series and systematic reviews. Q J Med. 2002; 95: 197–8.

21. Procopio M. Publication of case reports. Br J Psychiatry. 2005; 187:91.

22. Aggarwal R, Gogtay N, Kumar R, Sahni P, for the Indian Association of Medical Journal Editors. The revised guidelines of the Medical Council of India for academic promotions: Need for a rethink. Indian Pediatr. 2016; 53: 23–6.

PEARLS OF WISDOM

Preparing and Submitting Clinical Pictures

- The identifying features of the patient should be masked using permanent image editing, unless it is necessary to show the facial features.
- A written consent should be obtained from the patient/parents for publishing in the image in journal/book. A sample consent form is available at *www.indianpediatrics.net.*
- The photograph should be clicked with a high-resolution camera after ensuring sufficient light and its suitable direction.
- The relevant part should be sufficiently cleaned so as to avoid visualization of materials such as blood, pus, stains; unless it is necessary to show it.
- The image should be clicked against a lighter background giving it a contrast.
- It should be suitably cropped and should only highlight the part to be shown (Fig. 1) eliminating the unimportant part.
- There should not be any distracter in the image (e.g., jewellery, bright clothes with captions).
- The figure legend should be brief but sufficiently detailed to be understandable on its own.
- No abbreviations should be used in the figure legend, even if it has already been expanded once in the text.

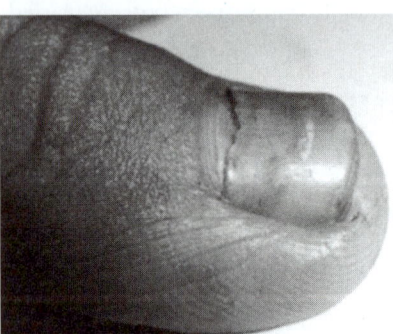

Fig. 1: Close-up view of nail of middle finger showing yellow discoloration with onychomadesis in a child with hand foot and mouth disease.
(*Image source:* Delayed cutaneous findings of hand, foot and mouth disease. Indian Pediatr. 2016;53:42–4.)

Shalu Jain and Dheeraj Shah

Preparing and Submitting Pathology/Microbiology Images

- **Figure 2** is a pathology image which is a clean merger of four different images.
- Each one of them is clearly labelled and shows only the necessary details.
- When two or more photographs are to be combined into one figure, each part of a composite figure should be clearly identified on the figure by letters (a, b, c, etc.)
- Selective enhancement or alteration of one part of an image is not acceptable.
- A good figure legend succinctly describes the content and enhances understanding with clear labels.
- Information about the figure (the figure legend) is placed directly below the image.
- Arrows should be used to highlight the important point. Letters, numbers, arrows, scales, and other marks that appear in a light area of the photo should be black.
- Stain used and magnification (×2, ×4.5, ×10, ×20, and ×40) should also be mentioned in the legend.

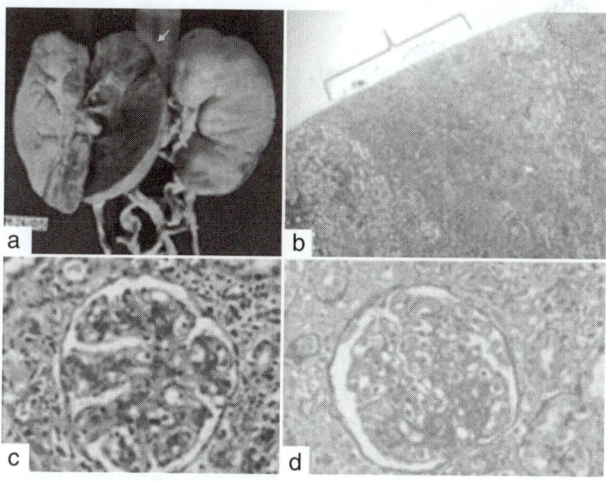

Fig. 2: (a) Kidneys were blotchy in surface and cut section showed discoloration and small infarcts (arrow); (b) Photomicrograph shows early infarct (marked); (c) and (d) Glomeruli only showed mesangial proliferation (H and E —b,c ,PAS —d, original magnification b— ×10, c and d— ×40).

(*Image source:* Necrotising polyarteritis nodosa-like vasculitis in a child with systemic lupus erythematosus. Indian Pediatr. 2017;54:139–44.)

Shalu Jain and Dheeraj Shah

Preparing and Submitting Radiological Images

- Ensure that the patient's name and identifying details are suitably masked by a permanent method of editing.
- Ensure that the image does not include any radio-opaque distracters (e.g., jewellery, markers, buttons).
- Part of the body that is imaged is to be written.
- The name of imaging study (e.g. X-ray, CT scan, MRI) should be clearly mentioned in the legend; do not leave it to the readers to find out.
- In cross-sectional imaging (CT, MRI, USG) section/plane/sequence/contrast-used is to be clearly mentioned; in X-rays, views to be mentioned.
- Arrows/markers should be used to direct readers' attention to the findings being shown in image (Fig. 3)
- Distorted images due to patient movement or images with any other kinds of artifacts should not be used.

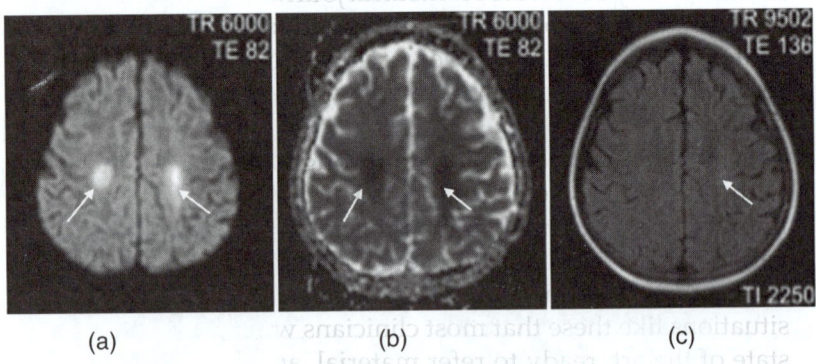

(a)   (b)   (c)

Fig. 3: Magnetic resonance imaging (MRI) of the brain in axial plane at the level of centrum semiovale in a patient during 1st episode leukoencephalopathy showing (a) high signal intensity in bilateral centrum semiovale on diffusion weighted imaging; (b) corresponding low signal intensity on ADC map; and (c) faint high intensity on fluid-attenuated inversion recovery (FLAIR) sequence. (*Image source:* Repeated episodes of leukoencephalopathy after high-dose methotrexate in a child with acute lymphoblastic leukemia. Indian Pediatr. 2017;54:159.)

Shalu Jain and Dheeraj Shah

11

Writing a Review Article: Making Sense of the Jumble

• Nitin Agarwal • Pooja Dewan

Medicine has witnessed a tremendous growth in recent times, and there has been a rapid emergence of new scientific evidence. Simultaneously, there has been a leap in the number of scientific journals. While there were about 6000 scientific journals in 1950, these increased to 28100 scholarly peer-reviewed journals in 2012; of these nearly 30% were biomedical journals[1]. With the availability of several sources of scientific information like scientific journals, conference proceedings, open archives, eBooks, and web pages, readers now are overwhelmed with a tsunami of scientific information.

While it would be ideal to go through all the literature dealing with the topic of concern and reach our own conclusions, it is impractical and virtually impossible to do so. Moreover, with large variations in clinical practice, young clinicians are often flummoxed as to which treatment or diagnostic strategy to adopt. It is in situations like these that most clinicians would seek some reliable, state of the art, ready to refer material, and a well-written review article would certainly fit the bill.

A review article is a comprehensive, critical analysis of published (and unpublished) material on a topic. In involves judicious and conscientious organization, integration, and analysis of the available literature relating to the topic of interest, to yield a summary which will help readers find a solution to the query. It helps to translate best evidence into best clinical practice. A review article serves as a useful guide for practising evidence-based medicine. Review articles provide a broad perspective of the problem at hand. They help to identify gaps, inconsistencies,

---

First published as Agarwal N, Dewan P. Writing a review article: Making sense of the jumble? Indian Pediatr. 2016; 53:715–20.

relations and contradictions in literature related to the topic of review. They apprise the readers of the state of current research while highlighting the avenues for future research. Cutting-edge reviews help clinicians to keep abreast with the latest developments. Box 11.1 summarizes the need for a review article[2-9].

NARRATIVE AND SYSTEMATIC REVIEWS

Review articles are classified as Narrative (or descriptive) and Systematic reviews. Typically, most review articles are written as narrative reviews, which are a summary of evidence derived from studies selected and interpreted according to the authors' personal review of literature. By providing comprehensive information on a topic, narrative reviews can help clinicians with even no or little knowledge of statistical methods or experimental designs to understand the cumulative scientific evidence regarding a clinical problem. Narrative review articles are particularly sought after by young researchers and students as they often provide them with not only a broad perspective of a clinical problem but also with solutions honed by years of clinical experience. However, narrative review articles also draw flak as they are prone to bias. Narrative reviews mostly reflect the authors' viewpoint based on their professional experience; while drafting a narrative review, authors

---

Box 11.1	Purpose of a review article

- To aid decision-making in clinical practice, e.g. diagnostic approach to primary immunodeficiency disorders[2].
- To decipher vexing problems on daily inpatient rounds, e.g. small for gestational age: Growth and puberty issues[3].
- To understand a sub-topic or question not part of conventional textbook e.g. massage and touch therapy in neonates: The current evidence[4].
- To summarize the enormous information available in a coherent and concise form, e.g. management of chemotherapy-induced nausea and vomiting[5].
- To identify gaps in current research, e.g. nutritional status of affluent Indian school children: What and how much do we know?[6]
- To identify relations, contradictions and inconsistencies in literature, e.g. treating hyperglycemia in the critically ill child: Is there enough evidence?[7]
- To identify emerging therapies, disease, or diagnostic aid, e.g. propranolol therapy for infantile hemangioma[8].
- To provide a direction for future research, e.g. immune thrombocytopenic purpura: Historical perspective, current status, recent advances and future directions[9].

may selectively include articles that support their hypothesis and exclude conflicting studies (selection bias). Synthesis bias can arise on account of the subjective approach of the authors while assimilating and synthesizing information while drafting a narrative review; the conclusion is often influenced by the author's personal opinion. Antman, et al.[10] found that narrative reviews often varied from the existing evidence, and were contradictory to other published expert opinions. The authors may limit their search of available literature to electronic databases, and freely available (full text) articles (search bias). Narrative reviews may also be marred by publication bias which may be due to tendency of journals to publish only studies with positive results (file drawer effect) or publish articles in English language. Some journals also give preferential treatment to publications from renowned investigators compared to lesser known researchers despite similar rigor in methodology. Funding agencies like pharmaceutical companies may also influence publication of sponsored research work, e.g. the company may not want to publish the results related to adverse effects of a drug marketed by them. Narrative reviews are characterized by the lack of an explicit description of the methods involved in research[10,11], and a lack of quantitative summary of the literature. In addition, narrative reviews are prone to plagiarism

Systematic reviews provide evidence-based synthesis of primary research studies in order to render an answer to a predefined research question[12,13]. A systematic review uses an explicit process to identify systematically and meticulously all studies pertaining to the specific research question, evaluates the methods of the studies, summarizes the results, presents key findings, discusses the reasons for variation in results between the studies, and analyzes the lacunae in current knowledge. The rigorous methodology of a systematic review helps to minimize the bias and ensures impartiality and reproducibility. Technically, systematic reviews are rated as the highest level of evidence by the US preventive services task force[13]. The process of a systematic review is laborious and can take years to complete. It is possible that the findings get outdated by the time it gets published with the emergence of new evidence. The cochrane systematic reviews are, therefore, much sought after as these are dynamic and updated regularly[14]. However, we must remember that systematic reviews can also be biased if the selection or emphasis of certain primary studies is influenced by the personal prejudices of the authors or funding sources. Also, as most of the systematic reviews address only a focused research question, it may not be possible to answer all questions related to that topic in one systematic review.

Table 11.1 summarizes the difference between narrative and systematic reviews. In this article, we aim to guide the readers mainly about narrative reviews, and how to make them objective and relevant. The approach to systematic reviews and meta-analyses is vastly different, and beyond the scope of this write-up.

WRITING A NARRATIVE REVIEW

Before Writing a Review Article

Scientific reviews are the most popular form of biomedical publication. Editors may wish to draw the interest of the readers by a review of a contemporary topic, either from the public health

Table 11.1: Salient features of types of review

	Narrative review	Systematic review
Topic/scope	Usually has a broader scope	Generally more specific and deals with a focussed research question
Appraisal	Qualitative appraisal often influenced by personal views of author	Critical qualitative and quantitative appraisal
Advantages	More popular among practicing physicians and young researchers as they can be understood easily without in depth knowledge of statistical methods and research methodology. They offer solutions to problems in question based on the experience and perspective of experienced authors.	Detailed and rigorous methods with predefined inclusion and exclusion criteria for primary studies. Clearly outlined search strategy. Lesser chances of bias.
Disadvantages	Not very rigorous methodology and results not replicable. More prone to bias.	Labour-intensive. Some knowledge of statistical methods is needed to understand it.

or research domain. The review article is also useful to disseminate the journal's perspective, or to provoke discussion regarding practice guidelines. Due to all the above reasons, most journal editors invite subject experts to write narrative reviews as the experience and stature of the expert increases the authenticity and readability of the article. The main author can be well-assisted by apprentices as co-authors, with mutual benefits, as a review article involves both in-depth knowledge and a painstaking study of the literature. This teamwork thus yields a much-savored product, useful to the biomedical community at large. However, not all review articles are solicited. In case you decide to write a review on a topic, it would be beneficial to get a go ahead from the editor of the journal you choose by sending him a proposal for the review along with your brief curriculum-vitae.

Choosing the Topic

It helps to write a review on a common clinical problem. A topic focused on a functional outcome, also referred to as Patient-oriented evidence that matters (POEM), rather than Disease-oriented outcome (DOE) would be preferred. For example, readers would be more interested in a review on drug A which improves the symptom score in patients with benign prostate hypertrophy rather than a review on drug B which merely increases urinary flow in benign prostate hypertrophy. Rapidly advancing fields require updated reviews. A new diagnostic test (e.g. Line probe assay for diagnosing pulmonary tuberculosis), emerging infections (e.g. Zika virus outbreak and its consequences) or new treatment modality (e.g. Caspofungin) are much sought after review topics. Any information that standard therapy is harmful is an equally good topic for a review article. It may be preferable to avoid reviews on very broad topics (e.g. hypertension in children). A narrative review which describes the etiology, pathogenesis, clinical features and management of a common clinical condition (e.g. headache) based on the author's interpretation and certain selected citations, may draw some favor from the general practitioners, but is not rated high from evidence-based perspective due to methodological flaws. Such broad topics may be better suited for chapters in textbooks. It also is preferable to avoid reviews on rarities or unusual manifestations of a disease, topics with only curiosity value and poor application, and a topic which lacks sufficient supporting evidence. Box 11.2 summarizes the points to consider while choosing a topic for writing a review article. We recommend that before you finalize the topic for a review article, you should get it approved from the journal you are planning to submit it.

Identifying the Research Question

The research question must be framed keeping in mind the contradictions, gaps, and inconsistencies in literature. The readers must be apprised of what is known and what your review aims to investigate. You must be able to formulate a clear, focused and relevant research question. You must be able to identify the five elements of the research question. These include the Population addressed, Intervention being evaluated, Comparator for the said intervention, the Outcomes being assessed, and Time frame (PICOT format).

Searching the Literature, Assessing the Quality of Literature and Integrating the Outcomes of Studies

Despite the unlimited information available online to both healthcare providers and the care seekers, the real challenge is to sieve through the haystack. Online health information may be surreptitiously advertising in content, providing outdated

Box 11.2	Topic selection for a review article

Topics to consider for a review article
- Specific illness/intervention/drug that concerns many readers
- Emerging health problem
- New drug/ vaccine/ diagnostic test
- Common clinical problems
- New guidelines for a condition
- Topic addressing patient-oriented outcomes (outcomes of importance to patients like changes in mortality, side-effects of a drug, etc.)
- Evidence that standard therapy is harmful

Topics to avoid for writing a review article
- Too broad
- Rarities or unusual manifestations of a disease; more suited for writing a case report
- Lack of sufficient supportive evidence
- Topic not suited to the journal you choose
- Topics with only curiosity value but poor appplication

Box 11.3	Check-list for assessing an online health resource

- Who is the publisher? A recognized, state-affiliated or international agency has greater authenticity eg. WHO, UNICEF
- What does it say?
  - Is the information plausible?
  - Does it have a reasonable hypothesis?
  - Is the information contemporary?
- What is the level of evidence and grade of recommendation?
- Does there appear to be a conflict-of-interest?
- Does the site appear to advertise or promote any health-related product?

information, misleading with respect to a drug or product, and/or, sponsored by an organization with inherent conflict-of-interest. Before citing an online health resource, a formal check-list (Box 11.3) is useful in avoiding later embarrassment.

Some of the credible sources of scientific information for a review article include journals indexed in medline, as most of them undergo a rigorous peer review process, and websites hosted by health organizations/associations of repute like world health organization, united nations children's emergency fund (UNICEF), centers for disease control and prevention (CDC), etc. Review articles based on reliable sources of evidence-based medicine like BMJ clinical evidence (*http://clinical evidence.bmj.com/x/index.html*), national guideline clearinghouse (NGC) (*http://www.guideline.gov*), and U.S. preventive services task force (*http://www. uspreventiveservicestaskforce.org/*) are given more weight. Websites like up to date offer a large variety of paid clinical reviews for practising clinicians; the funds are used to recruit experts, generally physicians in the field, as authors. The popularity of these is explicable as the topics are patient-oriented, contemporary, and updated[15].

Interpret the Evidence: Appraisal of the Retrieved Literature

After retrieving relevant literature, comes the more daunting task of reading and assimilating it. It is imperative to understand that all information may not be of the same standard or relevance for your narrative review. Some papers may be statistically weak (numbers too few or power too small), others may not have data that can be extrapolated to your scenario geographically or population-wise, and/or, some may have an inferior design (case series or case-control studies).

The most widely accepted hierarchy of evidence is the approach called Evidence-based medicine (EBM) (Box 11.4). Put simply, it implies that the most reliable information is obtained where there is minimum bias or sampling error. In other words, randomized trials or meta-analyses have been accorded the highest "Levels of evidence". The latter have undergone many variations since 1979[16,17]; the type of hierarchical table to be used is governed by the research question, whether it is prognostic, therapeutic, diagnostic or decision-making in nature. For example, in therapeutic studies, the highest level of evidence is attributed to a systematic review (of RCTs), the next is individual RCTs, while case-series and expert opinions are awarded bottom-place.

To enable clinical decision-making (and to enable an opinion in Narrative reviews), a system of graded recommendations has been in vogue since the early 2000s. This also has many modified forms, but largely it takes into account the number of RCTs or Level 1 evidence available, and also the consistency of the evidence. This ensures that all types of information are given appropriate credit[18]. We can thus understand that if the topic of the narrative review is relevant, and, if good evidence is unearthed and interpreted well by the author, the narrative review can be used to impart a strong recommendation or message to the readers. For example, if the review is about "current status of intravenous

---

Box 11.4 Levels of scientific evidence in decreasing order of merit[16]

IA   Evidence from meta-analysis of randomized controlled trials

IB   Evidence from at least one randomized controlled trial

IIA Evidence from at least one controlled study without randomization

IIB Evidence from at least one other type of quasi-experimental study

III Evidence from non-experimental descriptive studies, such as comparative studies, correlation studies, and case-control studies

IV   Evidence from expert committee reports or opinions or clinical experience of respected authorities, or both

Grades of recommendation[16]

A   Directly based on Level I evidence

B   Directly based on Level II evidence or extrapolated recommendations from Level I evidence

C   Directly based on Level III evidence or extrapolated recommendations from Level I or II evidence

D   Directly based on Level IV evidence or extrapolated recommendations from Level I, II, or III evidence

glutamine in malnourished children", and one finds two well-designed double-blinded RCTs along with three case-series from different countries all pointing to a similar result, the author is well-placed to provide a strong recommendation.

Present the Results/ Writing the Review

Since synthesizing literature and analysis of different opinions has a more intangible element to it than writing a standard research paper, beware of writer's block! It is also true that everyone has their own way of tackling this affliction of creative shutdown; the one way that works is to keep writing!

Before starting to write the main body of a narrative review, the following points need consideration. First, summarize your retrieved literature; editing these later can help you bring out your personal perspective and will minimize the chances of plagiarism. Second, brainstorming sessions with colleagues and co-authors will help raise angles and sub-plots to the main topic. These may drive a secondary literature search, ultimately improving the quality of the paper. It is important at this stage to include everything that you discussed; editing can be done later. Third, do not start writing with the introduction; keep this for the end, as it would provide you with a better 'bird's eye view' of the issue at hand. Fourth, divide your review into sections and subsections; these provide structure and flow and make it understandable. The aim is to organize the review like a story. For example, Devanarayana, *et al.*[19] have authored *"Recurrent abdominal pain in children"*. The sections and sub-sections used in this article are "epidemiology, clinical profile, etiology (organic and functional), management (pharmacological and non-pharmacological), public health perspective and prognosis." Clinical reviews can follow a similar structure as a practicing clinician can follow such sections with relative ease. Sections should follow each other logically and temporally. Fifth, to highlight and summarize important points in the review, make use of flowcharts, tables and boxes. It is prudent, however, to avoid duplication in text. In the above-mentioned example, a useful box would be 'Recent radiological investigations for recurrent pediatric abdominal pain'. Tables can depict the comparative results of different studies included in the review. Construct tables with studies in rows, while columns should indicate the proposed characteristics. For example, a review on

'Portal hypertension and hydatid cysts in children' could include a table of reported case series, wherein the columns could show 'Number of cases, Country of Study, Type of portal hypertension (hepatic or post-hepatic), Treatment given, and Mortality.'

Finalizing the Narrative Review

As mentioned earlier, complete the introduction now; keep in mind the background knowledge on the subject and the lacunae in literature, the target audience, your findings from literature, and your proposed recommendations. About 200–300 words can effectively convey the feel of the coming text. After this, a second and third appraisal of the entire manuscript is valuable for eliminating errors and possible plagiarism. Then, finalize and check all references.

Avoid plagiarism: Plagiarism is the deliberate or inadvertent copying of words, phrases, data, ideas or figures and claiming them as your own. It is the most recognized unethical practice since it violates a basic tenet, i.e. honesty, while science is essentially a search for truth. Avoiding plagiarism should be a key consideration for any biomedical writer as it can be a major source of embarrassment and/or censure/ blacklisting for the author and the journal. In review articles, a lot of information, opinions and results are studied, tabulated, and analyzed. There is genuine scope for inadvertent plagiarism creeping into the manuscript; technology has, ironically, been culpable in this regard[20]. The same technology; however, can backfire, and pick up direct plagiarism of text, or indirect re-hashing of ideas (which has been attempted as a cover-up). Box 11.5 guides the authors on how to avoid plagiarism.

Box 11.5	Tips to avoid plagiarism

- Read and understand all the subject matter thoroughly; do not rely only on abstracts
- If there is the slightest chance that the idea behind your review has been addressed before, it is better to acknowledge the source in advance
- Understand, contextualize the information and reproduce
- Use references liberally, except when the fact is commonly known to all
- Avoid using downloaded or previously printed images or charts
- Use appropriate format for references
- Use software like thenticate, cross check, plagiarism checker to proofread your manuscript before submission

CONCLUSIONS

A review article is an important source of information for evidence-based medicine. It serves as a ready to use reference for all health professionals. A review must address a clinically relevant issue with significant implications for patient-care. The topic for review should be relevant, contemporary, and deal with a focused research question. A good review must be rigorous, up-to-date, and unbiased. The conclusions of the review must be well supported by the analysis of literature and should include: summary of the present problem, clinical practice guideline or recommendation depending on the level of evidence you have unearthed, and/or, directions for future research to fill gaps in existing literature.

REFERENCES

1. International association of scientific, technical and medical publishers. The STM Report: An overview of scientific and scholarly journal publishing. Netherlands: International association of scientific, technical and medical publishers; 2012. p.110.
2. Madkaikar M, Mishra A, Ghosh K. Diagnostic approach to primary immunodeficiency disorders. Indian Pediatr. 2013; 50: 579–86.
3. Yadav S, Rustogi D. Small for gestational age: growth and puberty issues. Indian Pediatr. 2015 ;52: 135–40.
4. Kulkarni A, Kaushik JS, Gupta P, Sharma H, Agrawal RK. Massage and touch therapy in neonates: the current evidence. Indian Pediatr. 2010;47: 771–6.
5. Dewan P, Singhal S, Harit D. Management of chemotherapy-induced nausea and vomiting. Indian Pediatr. 2010;47: 149–55.
6. Srihari G, Eilander A, Muthayya S, Kurpad AV, Seshadri S. Nutritional status of affluent Indian school children: what and how much do we know? Indian Pediatr. 2007;44: 204–13.
7. Poddar B. Treating hyperglycemia in the critically ill child: is there enough evidence? Indian Pediatr. 2011;48: 531–6.
8. Gunturi N, Ramgopal S, Balagopal S, Scott JX. Propranolol therapy for infantile hemangioma. Indian Pediatr. 2013;50: 307–13.
9. Anoop P. Immune thrombocytopenic purpura: historical perspective, current status, recent advances and future directions. Indian Pediatr. 2012;49:811-8.
10. Antman EM, Lau J, Kupelnick B, Mosteller F, Chalmers TC. A comparison of results of meta-analyses of randomized control trials and recommendations of clinical experts. Treatments for myocardial infarction. JAMA. 1992;268: 240–48.

11. Bramwell VH, Williams CJ. Do authors of review articles use systematic methods to identify, assess and synthesize information? Ann Oncol. 1997;8: 1185–95.

12. Garg AX, Hackam D, Tonelli M. Systematic review and meta-analysis: when one study is just not enough. Clin J Am Soc Nephrol. 2008;3: 253–60.

13. Montori VM, Wilczynski NL, Morgan D, Haynes RB; Hedges Team. Systematic reviews: a cross-sectional study of location and citation counts. BMC Med. 2003;1:2.

14. Cochrane. Trusted evidence. Informed decisions. Better health. Available from: *http://www.cochrane.org/about-us*. Accessed June 21, 2016.

15. UpToDate®. Available from: *http://www.uptodate.com/home/about-us*. Accessed July 16, 2016.

16. National Guideline Clearinghouse. Levels of evidence and grades of recommendation. Available from: *https://hsl.lib.umn.edu/biomed/help/levels-evidence-and-grades-recommendations*. Accessed July 4, 2016.

17. The periodic health examination. Canadian Task Force on the Periodic Health Examination. Can Med Assoc J. 1979;121: 1193–1254.

18. Burns PB, Rohrich RJ, Chung KC. The levels of evidence and their role in evidence-based medicine. Plast Reconstr Surg. 2011; 128: 305–10.

19. Devanarayana NM, Rajindrajith S, De Silva HJ. Recurrent abdominal pain in children. Indian Pediatr. 2009; 46: 389–99.

20. Li Y. 'Standing on the shoulders of giants': Recontextualization in writing from sources. Sci Eng Ethics. 2015; 21: 1297–314.

12

Responding to Reviewers' Comments

•*Peeyush Jain* •*AK Patwari*

You have finally submitted your manuscript to a journal of your choice. Now that the proverbial elephant has been pushed through the door, a heavy load has been removed from your shoulders and you are eagerly awaiting the elusive acceptance letter from the editor. But beware; the elephant's tail is going to get stuck more often than not! It is incredibly rare for a manuscript to be accepted without revision on first submission.

MANUSCRIPT HANDLING AT THE JOURNAL OFFICE

On submission of the manuscript, the editorial review process begins with an initial reading by the Chief editor, looking at the relevance and novelty of the manuscript and its conformity to the journal guidelines, before either rejecting or considering it suitable for journal review process. Once considered suitable, it is then forwarded to reviewers or to one of the associate editors to assigns reviewers, who are subject experts in a field related to submitted manuscript. The reviewers report back within a given time frame with their comments, both for the authors and confidential comments for the editors. The editor then uses these reviews to make a decision on suitability of the manuscript for publication in the journal. The comments of the editors and the reviewers are then conveyed to the corresponding author.

There are four primary types of editorial decisions that are made: Acceptance, Minor revisions needed, Major revisions required, and Rejection—with the first one being highly uncommon. Any author worth his salt would have received most of these editorial decisions in his writing career.

First published as Jain P, Patwari AK. Responding to Reviewers' Comments? Indian Pediatr. 2016; 53:1093–95.

If revisions are required, the author is required to resubmit the manuscript after modifying in light of the comments. The editor would again go through it and in all probability forward it to the same reviewers to assess whether their comments have been addressed satisfactorily and seek their opinion about suitability of publication of the revised manuscript. At any step, the editor and/or the reviewers may return the manuscript with a request for further revisions. The process continues till the modifications have been made to the satisfaction of the reviewers and the editor. The acceptance of the manuscript is then conveyed to the corresponding author.

PREPARING THE RESPONSE

There are slim chances that your manuscript may be accepted without any changes. If this happens, you may count yourself lucky, because such an editorial response is not easily forthcoming[1].

If only minor revisions have been suggested, it is advisable to do so without making any fuss. You should then send the revised manuscript back to the editor as soon as possible.

Request for major revisions, is the commonest editorial decision made. This means that the current version of your manuscript has been rejected, but would be reconsidered after the suggested changes have been made. It may feel like a body blow even to seasoned authors. But you must always remember that if major revisions have been suggested, you are still in the game and have a good chance of success. Moreover, the manuscript under evaluation is a product of your hard work. You can not afford to let it go without 'fighting' it out with all your might.

Once rejected, there is little you can do to get your paper published in the same journal. You must revise your manuscript according to the suggestions received and submit it to another suitable journal for publication.

An editorial decision for changes usually indicates that the Journal is interested in your article, which is good news[2]. In any other case, one must not be discouraged and you would be well advised not to fret. Take a break, and then re-read the comments. One should not take the reviewers comments personally. In most cases, the reviewers are blinded to the authors' identity and the place of study. You must realize that they are only doing their job to the best of their ability and that too on complimentary basis just for love of research. Editors strive hard to make their journal better and this they can achieve by helping you improve your manuscript. Even though the comments may appear grave and direct; by and large they have been made in the spirit of constructive criticism[3]. After the cooling-off period, discuss the editorial decision with your co-authors and get to work on it. As all authors are equally responsible for the paper, it is prudent to get their observations and suggestions on the comments received and approval on the changes to be made.

It is worthwhile for an author not to immediately react to the comments of the reviewer with defensive reasoning. Try to understand the reasons behind the comments, look at the issues with a critical eye just like an 'editor' and not as an author. This will help you appreciate the reviewer's point of view and may even help you to identify other factual or typographical errors which might have been previously missed. After having 'educated' yourself regarding the deficiencies in the manuscript you would find yourself in a better condition to revise the manuscript and respond to the reviewers comments.

REVISING THE MANUSCRIPT

Based on the reviewer's comments, make the changes in the manuscript as planned. It is prudent to adhere to the journal guidelines while formatting the manuscript. Underlining or highlighting the modifications in a different color will make them easily visible. If the journal so accepts, the 'track changes' option of the software could be used to highlight the revision done. All this will help to facilitate the review. Do remember to also send a 'clean' (without track changes and highlighting) copy. The referencing style and the in-text citation must also be re-checked after revision.

Most of the Journals either use 'US English' or 'UK English" as their preferred choice of language; check the journal guidelines and set your word processor accordingly. If there are major grammatical mistakes the reviewers may want you to get the manuscript vetted by a person who has reasonable expertise in the English language. You may sometimes be asked to reduce the size of your manuscript. Deletion and/or mergers of tables and/or figures is also requested by the reviewers. Suggestions of inserting of bulleted lists may also sometimes be given.

As such these reviewer suggestions, especially those which lead to reduced word-count of the manuscript, should ordinarily be accepted. If the authors feel that suggested changes are not advisable, the editors should be requested for reconsideration. Journals are usually short of printing space and editors have to keep the final layout in mind, and so chances are that the editor is unlikely to heed to such requests. The other option of sending the unrevised manuscript to a different journal is always available.

Authors may consider a complete rewrite of the manuscript incorporating the suggestions available. But unless requested to do so by the reviewers, it is likely to be considered a new submission and the whole rigmarole and the emotional roller coaster ride of manuscript submission process would have to be faced again. In any case, treat the revision with same diligence as you had shown to the original version. Do not abandon the revision and the subsequent resubmission for long, as the chances of acceptance decreases with time

You may decide not to make any changes in the manuscript and submit the same to some other journal. But here lies the catch! There are chances that the manuscript may land up with the same reviewer as before. In that situation the reviewer is unlikely to bestow any kindness if his earlier issues have not been addressed to his satisfaction. As such any inherent issues present in the paper would have to be sorted out before the manuscript is finally accepted.

RESPONDING TO COMMENTS

The response process is akin to be interviewed, you would be required to defend your position and paper. Do not underestimate the task. It might sometimes take a long time and a lot of effort to address the issues

The revised manuscript should be accompanied by a covering letter detailing point-wise modifications that have been made in the manuscript. The journal guidelines should be adhered to in the layout of the letter. There are three basic 'tips' or golden rules for responding to reviewer's comments. Answer politely, answer completely, and answer with evidence[4]. It would be a good gesture to begin the letter by thanking the reviewers for their effort and time spent. A gratuitous approval in acknowledging the same is always appreciated by the reviewers and editors alike. The information given should be clear and concise. Do give raw data when asked for. The reviewer may want to cross check your results or calculations.

The comments of the reviewers/editorial board should be typed verbatim (ideally use copy-paste commands of the word processor) and point-by-point response provided while giving sufficient attention to details. If the comments are too many, they may be depicted in a tabular form with comments on the left and your responses on the right side. Try to identify the main concerns of the reviewers. Even if the comments are in form of big paragraphs it is advisable to split them into separate points and address them individually[1]. The changes done in the revised manuscript need to be mentioned along with the place where they have been made. The numbers of the page and the lines therein may also be cited wherever possible.

It is suitable to use present tense or past perfect tense in the covering letter, to intimate the changes made; for example" We now add the following sentence..../ we have added the following line ...."[2]

It is not always feasible for the Journal editor to get two subject experts to review the manuscript within stipulated time. So he/she may have to get the manuscript reviewed by general experts only. On such occasions the comments from the reviewers may be conflicting and sometimes even differ drastically; though such occurrence is rare, it would be reasonable to go with one reviewer's comments, justifying your reason to do so.

It is also human nature to make mistakes. Reviewers and even editors are known to make their share of errors in the review process[3]. You might sometimes not be willing to accept the opinion of the reviewers and are of the opinion that he/she is wrong. There is always an option of 'disagreeing' with the reviewer on certain points or comments. You would need to be thorough in your reply and give well-supported reasons for your divergent views. You will be well advised to watch your tone and tenor in your reply. Do not try to justify/argue unnecessarily. After all you need the reviewer/s to only yield to your point of view rather than become your ardent supporter.

You can always write to the editor imploring him to change the decision. However deriding the reviewers and questioning their judgment is a low yield strategy[5]. It is more in your interest to give it another shot and get your work published rather than to ignore it[6].

The Road Ahead

Even if your paper has been rejected, it is not the end of the world. In a previous study[7], the top reasons for rejection of articles submitted to Indian Pediatrics in one calender year were 'absence of a message', 'lack of originality', 'inadequate methods', 'not relevant to journal', 'over-interpretation of results', 'unsatisfactory writing style', 'inaccurate/inconsistent/insufficient data', and 'inappropriate statistical analysis', in that order. The rejection may not be a reflection of your manuscript. It happens, that for stated or unstated reasons, the editors decided that your paper was not what they wanted. It may just have been bad timings as the journal might have just published or accepted a study similar to yours[5].

A good paper would always get published in one or the other journal more often than not. In fact, out of papers rejected by *Indian Pediatrics* in the year 2002, 18% were detected to have been subsequently published elsewhere, some even in journals with Impact Factor higher than that of the rejecting journal[6].

The Essence of Response to Reviewers

One should consider the reviewers' comments as free expert advice on the shortcomings of the manuscript. You must take the comments with all seriousness and respond to best of your ability. After all; you, the editors and the reviewers all are on the same side and would like to see a quality manuscript getting published.

REFERENCES

1. Rocco TS, Hatcher TG, editors. The Handbook of Scholarly Writing and Publishing. San Francisco: Jossey Bass; 2011.
2. Kotz D, Cals JWL. Effective writing and publishing scientific papers, part XII: responding to reviewers. J Clin Epidemiol. 2014;67: 243
3. Shaw JD. Responding to reviewers. Acad Manage J. 2012;55: 1261–63.
4. Millette D. Dealing with reviewer's comments. J Orthod. 2006;33: 69–70.
5. Cummings P, Rivara FP. Responding to reviewers comments on submitted articles. Arch Pediatr Adolesc Med. 2002;156: 105–7.
6. Dewan P, Gupta P, Shah D. Fate of articles rejected by Indian Pediatrics. Indian Pediatr. 2010;47: 1031–35.
7. Gupta P, Kaur G, Sharma B, Shah D, Choudhury P. What is submitted and what gets accepted in Indian Pediatrics: Analysis of submission, review process, decision making, and criterion for rejection. Indian Pediatr. 2006;43: 479–89.

Blowing Your Own Trumpet: How to Increase the Online Visibility of Your Publication

• *Sumaira Khalil* • *Devendra Mishra* • *Deepika Upadhyay*

After sustained efforts leading to the completion of the research-work and its acceptance and publication in a journal, the authors heave a sigh of relief and plan to enjoy the moment. However, be aware that some more effort is still needed to ensure that the publication receives the attention of other workers in the field and gets maximum visibility. Millions of research papers are published every year[1], both in print and on the Web finding ways of increasing the visibility of your scientific work is, thus, of utmost importance.

With increasing access to the internet, avenues for dissemination and accessibility of research have widened significantly. Online portals and websites are increasingly becoming a tool for promoting one's research publications to attract more readers and professional colleagues–thereby expanding your professional network. We herein discuss some of the common ways to enhance the visibility of an accepted/published research paper among other researchers and also lay persons. The cornerstone of this strategy is to be aware of this need for wider dissemination early on, and start working for it from the stage of article preparation itself.

DURING MANUSCRIPT PREPARATION

Choosing the Journal

Choosing the right journal for publishing your paper is the first step towards reaching your target audience. This important decision must be made before you start writing your article, and has already been discussed in a previous chapter (*see* page 5)[2]. Springer's journal suggester (*www.journalsuggester.springer. com*),

First published as Khalil S, Mishra D, Upadhyay D. Blowing your own trumpet: How to increase the online visibility of your publication? Indian Pediatr. 2016; 55:49–54.

Elsevier's journal finder (*www.journalfinder.elsevier.com*) and Journal guide (*www.journalguide.com*) are some tools available online that will help you to choose the most suitable journal for your research. Find my journal (*www. findmyjournal.com*) is a free-software that indexes all major publishers and provides a wide database to search the most appropriate journal to publish your research work, once you provide some information about your paper.

Open-access journals: Research shows that citations of a published paper increases by 10% if published in an Open access journal[2,3]. Open access (OA) journals are online journals with free subscription where the cost is incurred by the author for publishing, and it gives full online access of all the articles. Articles published in open access journals are available free full-text to all readers. Some limitations of OA journals are the increasing prevalence of ghost journals and predatory journals[4], which has already been addressed in detail[2].

Indexed journals: Articles submitted in indexed journals are considered to have higher scientific quality as compared to non-indexed journals, thereby increasing their readership. There are numerous indexing services like EMBASE, SCOPUS, SCIRUS, Medline and EBSCO Publishing electronic databases[5], and authors need to ensure that the journal is indexed in a reputed and acceptable database.

E-publication ahead of print: Some journals have a policy of publishing accepted papers on their website, prior to their scheduling for the print issue. These are posted to the website for making them available to readers, ahead of their publication in print (though the text may undergo some changes in the final version after proof-reading). Publication in such journals makes your accepted manuscript available online very soon after acceptance, thereby precluding the delay till print issue.

Title and Abstract

The title and abstract of an article are most commonly used by the indexing systems and databases for conducting literature search, as has already been detailed[6]. Writing a search engine-friendly title and Abstract can be one of the ways of ensuring the visibility of your paper to a wider audience[6]. Thus, write a concise and interesting description (50–100 words) emphasizing the prudent points, and anything new added to the existing knowledge as Abstract. Structured abstracts are preferred by most journals but various innovations have also come up in recent times:

Graphical abstracts: This is a *"single, concise, pictorial and visual summary of the main findings"*[7]. These graphical abstracts can be used for promoting your article by tweeting it, sharing on social media, or adding a link to your article. Detailed information regarding graphical abstracts is available at various journal websites.

Video abstracts: Some journals also ask for a video abstract, linked to the e-version of the paper once they are accepted for publication[7]. These are about 5-minute videos describing the main aim and results of the paper using animation and graphical representation. Tips to prepare a video abstract are also available at the author services pages of many publishers.

Tweetable abstracts: Some journals, apart from traditional abstracts, also ask for a tweetable abstract[7], comprising of a summary of the aim and results of the study in about 100 words. This facilitates rapid dissemination of information to a wider audience.

AFTER PUBLICATION

After the publication of the article, the authors' efforts should be directed to ensure that it reaches the widest possible (and most appropriate) audiences. This can be done by making it available on the Web, and sharing information about it through social networks and professional networks.

Using the World Wide Web

There are innumerable opportunities on the Web for disseminating information about your published paper, and sharing it with interested professionals (Box 13.1).

Using personal e-mail: You could very well include your article in your email signature. Some journals even provide a free service for preparing a banner related to your paper, so that it can be placed at the bottom of all your e-mails. As many of the persons you correspond by e-mail are also working in the same field, this method is most likely to reach a relevant audience. In addition, a short message (with/without a link) can also be posted to any discussion lists or Listserves you are a member of.

Box 13.1	Highlighting your article on the web

- Talk about your research on social media e.g., blogs, Facebook.
- Share your e-print link with your networks.
- Add a link to your article on your email signature.
- Add your article to reading lists.
- Post a link to your article on your profile on all professional or academic networks you're a member of.
- Put a link to your article on your institutional profile page, your personal webpage or any project websites.

Online repositories: These are created by single institutions, cross-institutional or departmental; where members of the institution can deposit their research work; which is freely accessible. Institutional repositories preserve and disseminate the research output of the institution in the form of digital copies of major research projects, dissertations, published papers of faculty, etc. at the institutional website for free access by other researchers. The contents available are mainly theses, unpublished research projects, pre-prints, post-prints and research reports and does not undergo a peer review process e.g., D space is a repository created by the Massachusetts Institute of Technology to manage and disseminate all the research and intellectual output from the university[8].

Even if your departmental/institutional website does not currently have an archiving system, you can still use the institutional website to embed links to your published paper.

Social science research network (SSRN): SSRN is a popular website which provides for dissemination of research papers and scholarly work in the area of social sciences and humanities. It is an open access repository for the scholarly community wherein articles/papers can be uploaded for free access by those who have an account on the website.

The functioning of open access repositories can be monitored by Registry of open access repository (ROAR) and Open directory of open access repository (open DOAR)[9]. One should utilize repositories to make available their published paper (even if it is only in the abstract format), as this practice leads to a wider access to the research, thereby increasing its visibility. However, before sharing the full-text of your published paper on any such website, do check with the journal where it is published–many journals do not permit archiving of full-text articles, not even self-archiving!

Self-archiving: One of the ways of self-archiving accepted article is through creation of personal websites that makes the article open access, thereby increasing its impact and citations. Personal websites are created by individual researchers where they post information regarding their article along with their personal details, a curriculum vitae, list of publications, research interest, experience and expertise, and research projects and conference presentations[10]; thereby, making it a channel that projects their work. It has been found that open access repositories have a lower proportion of citations[11] as compared to personal websites, but the effectiveness of such self-publishing is questionable[12]. There could be varied reasons for creating personal websites, for expression of identity, self-presentation or even entertainment quotient[13,14]. Again, the issue of copyright crops up, as many journals do not permit this self-archiving of the published paper (and they hold the copyright!).

Search engine optimization (SEO): Web-based search engines like Google scholar, Microsoft academic search, Elsevier's scirus, SciDiver, IEEE Xplore, SciPlore.org and others are commonly used for searching academic articles. These search engines index PDF files of academic articles making it freely accessible online, thereby increasing its visibility[15]. However, with the presence of web spams, the citation counts on google scholar must be taken with a pinch of salt. By optimizing your articles, you guarantee that your articles are indexed and gain a higher ranking in general and academic search engines. SEO of an article helps to keep your article on the top when someone runs a related search on search engines like Google. This leads to greater visibility amongst readers and potentially more citations[16].

There are several ways you can optimize your article for better indexing and ranking in search engines. Using strong keywords and synonyms (preferably incorporating Medical Subject headings from PubMed), the location in the text of these keywords (i.e., in a title vs. only in a sub-heading), the completeness of the metadata, the use of vector graphics for your graphs, and having a public group for your research, are some examples of the techniques[17].

Using the Social Network

With the extensive use of internet, social networking has become one of the most common media for projecting your research work (Box 13.2). Many researchers post their work on Facebook, Twitter, WECHAT, etc. for widespread visualization, awareness, to start

Box 13.2   Writing about your research on the web
- Remember, non-professionals and general audience may also read your post.
- Keep the language as simple and jargon-free as possible.
- Avoid over-simplification of study-implications.
- Avoid inflating the importance of yours findings.
- Provide links to supporting material and your institutional/personal websites.
- Tag friends, colleagues, and other researchers in the same field.
- Provide figures /illustrations /pictures; they generate interest.

interesting discussions and increasing visibility. Blogs are also an important and very effective way of disseminating information about published research. Using the social media, in addition to disseminating your research to your social circle, is also a good way to reach the lay public, interested patients, and even media personnel, sometimes leading to mention in the media publications, if it is a newsworthy research.

There are multiple scientific social networks like Research gate, Linkedin, Mendeley and Academia.edu. In these academic networking sites, apart from projecting your research work, you can also build a complete picture of your professional achievements and research interests.

Research gate: Research gate is one of the more popular scientific social networking website where you can read and share publications, create personal profile for sharing your research work from the project-stage till they are accepted for publications, and is an excellent platform for generating discussions[18] and connecting with other academicians; thereby, making your research visible. Researchers must have a user ID with the website through which they can freely upload their articles/working papers/conference papers and scholarly works for free access by other users. Research gate also has its own citation impact measurement in form of RG Score. In recent years, it has been noted that many article published in 'ghost' or 'predatory journals' are also visible on research gate, highlighting the poor reliability of such websites[19].

SciENcv (science experts network curriculum vitae): The National Center for Biotechnology Information (NCBI), in addition to the PubMed, also provides a researcher profile system for researchers who apply for, receive or are associated with research investments from US federal agencies. SciENcv is available at My NCBI on the NCBI website (*https://www.ncbi.nlm.nih.gov/sciencv/*).

Atlas of science: This is another free website that aims to disseminate scientific information, especially the researcher's own perspective on the results, to the general public at a faster speed. This, by-invitation service, permits the researcher to write and publish a short lay summary of their peer-reviewed published article (including illustrations), which is then available at the Atlas of Science website[20]. The service is free to both the authors and the readers.

Kudos is another free web-based service that helps researchers increase the visibility and impact of their publications, in addition to providing the facility of providing short lay person summary of research for sharing with various website. Various article metrics are also provided (*https://www.growkudos.com/about*).

Using the Journal's Services

Once a manuscript is accepted by a journal, the journal has an almost equal stake in the increased visibility and citation of the article. Thus, most journals have developed various mechanisms for increasing the visibility and accessibility of the published article, and their help pages guide authors on utilizing these. Some of these are:

E-prints: Providing e-prints of the published paper is a method of increasing visibility of accepted articles e.g., authors of any article published in a taylor and francis journal get a limited number of free e-prints to share with their colleagues in order to enhance its visibility[21]. E-prints can also be shared by linking it to one's e-mail or posting it on one's twitter account or Facebook; the link provided by the publishers continues to work after the free access allowance is reached, but then it directs people to the abstract page. Another example is the springer nature content sharing initiative using a sharedIt link, wherein additionally, the readers are also able to use enhanced PDF features such as annotation tools, one-click supplements, citation file exports and article metrics[22]. Some journals even provide this free e-offprint along with the final cover of the respective issue, making it more attractive for sharing.

Persistent identifiers: These are unique numbers/alphanumeric identifiers for scholarly works, either a journal (e.g., ISBN) or an individual article (e.g., PMID used in PubMed). These are *"enduring references to a resource such as a web page, file, image, or other (usually digital) object"*[23]. Persistent identifiers can also be used to locate either the resource or information about the resource–usually by using it as a URL. A digital object identifier (DOI) is one such adapted identifier system that provides alpha-numeric identifiers that can be turned into a URL. However, it is a paid system, whereas the publisher item identifier (PII) is another one which is free to use by journals, and is also used by a number of scientific journal publishers[24]. It uses the pre-existing ISSN or ISBN of the publication in question, and adds a character for source publication type, an item number, and a check digit.

Persistent identifiers can also be obtained by individual researchers (unique researcher identifiers). Open researcher and contributor ID (ORCID) links to an individual's research output and helps in distinguishing researchers from each other[25], and is being increasingly used by journals, institutions and recruiters. Other such identifiers include researcher ID (integrated with web of science) and scopus ID.

Direct link to social media: Some journals provide a direct link to facebook/twitter/pinterest etc. directly from your published paper at their website. Thus, the author is saved the hassle of writing a message and embedding the link to the article at these sites—all this is available readymade at the click of a button.

Delineated contributor roles: Although not strictly a method to increase visibility of the article, detailing standardized contributor roles with the article increases the visibility of the individual authors, especially in relation to their core competence with respect to the publication. Some journals, e.g., medicine, have integrated CRediT (contributor roles taxonomy) into their submission system, enabling more transparency to the published work and allowing individual authors to get credit for their specific contributions to the manuscript. The system provides a list of 14 author's contribution roles for selection with each author's name, with more than one contribution permitted for each author[26].

Over the years, the pressure on researchers is not just limited to publishing the research, it has gone on to creating an impact and demonstrating that impact (by citation metrics etc.). With the availability of online media and social network, there are numerous tools available to make your presence felt and increase the visibility of your research, which indirectly increases your citations and helps in expanding your professional network. But these methods also come with their share of drawbacks in the form of ghost journals and web spams. We have discussed a good selection of the armamentarium at your disposal (Box 13.3), though new ones

---

| **Box 13.3** | **Increasing a publication's visibility** |

During manuscript preparation
- Choosing the journal
  - Best match for your subject-area
  - Indexed journal, preferably with e-publication ahead of print
- Title & Abstract
  - Write a search-engine-friendly title and abstract
  - Use facility for graphical/video/tweetable abstracts

After publication
- Using the web
  - Personal e-mail: Share, use article link as a mail header
  - Online repositories: Upload abstract/full text at such repositories
  - Self-archiving: Upload link at your personal, or institutional website
  - Search engine optimization: Optimize your article for identification by search engines
- Using the social network: Use your social network to discuss your research.
- Using the journal's services: Enquire about and utilize the journal's author services for increasing your publication's visibility.

---

would be available by the time we are in print; it is up to the authors to remain updated. The right balance between good scientific vigor in research and awareness of modalities to disseminate it will help a researcher in making a stronger academic reputation in the future.

REFERENCES

1. Jinha AE. Article 50 million: An estimate of the number of scholarly articles in existence. Learned Publishing. 2010;23: 258–63.

2. Dewan P, Shah D. A writer's dilemma: where to publish and where not to? Indian Pediatr. 2016;53: 141–5.

3. Jayaprakash K, Rekha AP, Rajendiran S. Open access journals-A study. IOSR J Human Soc Sci. 2013;8: 7–9.

4. Mirjana I, Miloš R, Vladimir K, Jovana V. Computer science and information systems: Publishing an international open access journal in a developing country. Proceedings of the 5th Belgrade International Open Access Conference, Belgrade; 2012. p. 101–12. Available from: *http://boac.ceon.rs/public/full/5th-bioac.pdf*. Accessed October 1, 2017.

5. Balhara YPS. Indexed journal: What does it mean? Lung India. 2012;29:193.

6. Dewan P, Gupta P. Writing the title, abstract and introduction: looks matter! Indian Pediatr. 2016;53: 235–41.

7. Hartley J. What's new in abstracts of science articles? J Med Libr Assoc. 2016;104: 235–7.

8. Ware M. Universities' own electronic repositories yet to impact on Open Access. Nature Web Focus. Available from: *http://www.nature.com/ nature/focus/accessdebate/4.html?foxtrotcallback=true.* Accessed October 1, 2017.

9. Zainab AN. Open access repositories and journals for visibility: Implications for Malaysian libraries. Malaysian J Libr Inform Sci. 2010;15: 97–119.

10. Bleda AM, Aguillo IF. Can a personal website be useful as an information source to assess individual scientists? The case of European highly cited researchers. Scientometrics. 2013;96: 51–67.

11. Chen C, Sun K, Wu G, Tang Q, Qin J, Chiu K, et al. The impact of internet resources on scholarly communication: A citation analysis. Scientometrics. 2009;81: 459–74.

12. Barjak, F. The role of the Internet in informal scholarly communication. J Amer Soc Inform Sci Technol. 2006;57: 1350–67.

13. Marcus B, Machilek F, Schütz A. Personality in cyberspace: Personal web sites as media for personality expressions and impressions. J Personality Soc Psychol. 2006;90: 1014–31.

14. Weibel D, Wissmath B, Zroner R. Motives for creating a private website and personality of personal homepage owners in terms of extraversion and heuristic orientation. Cyberpsychology. 2010;4:5. Available at: *https:/ /cyberpsychology.eu/article/view/4234/ 3278.* Accessed September 30, 2017.

15. Anonymous. Get Noticed: Promoting Your Article for Maximum Impact. Elsevier. 2016. Available from: *https://www.elsevier.com/__data/assets/ pdf_file/0013/201325/Get-Noticed_Brochure_Sep2016.pdf.* Accessed October 1, 2017.

16. Beel J, Gipp B, Wilde E. Academic Search Engine Optimization (ASEO): Optimizing scholarly literature for Google scholar and co. J Scholarly Publ. 2010;41: 176–90.

17. Anonymous. Get Found. Optimize Your Research Articles for Search Engines. Elsevier, 2016. Available from: *https://www.elsevier.com/connect/ get-found-optimize-your-research-articles-for search-engines.* Accessed October 1, 2017.

18. Thelwall M, Kousha K. ResearchGate articles: Age, discipline, audience size and impact. J Assoc Information Sci Technol. 2017;68: 468–79.

19. Memon AR. Research Gate is no longer reliable: Leniency towards ghost journals may decrease its impact on the scientific community. J Pakistan Med Assoc. 2016;66: 1643–47.

20. Atlas of Science. About Us. Available from: *http://atlasof science.org/sample-page/.* Accessed October 1, 2017.

21. Author Services. Share your e-prints. Available from: *http://authorservices. taylorandfrancis.com/shareyour eprints/.* Accessed October 1, 2017.

22. Springer Nature. Shared. Available from: *http://www. sprin gernature. com/ gp/researchers/sharedit.* Accessed October 1, 2017.

23. Anonymous. Database Linking. *https://www.elsevier.com/authors/author-services/.../data-base-linking.* Accessed October 1, 2017.

24. World's Veterinary Journal. Publisher Item Identifier. Available from: *http://www.wvj. science-line.com/ wvj-pii.html.* Accessed October 1, 2017.

25. Open Researcher and Contributor ID. What is ORCID? Available from: *https://orcid.org/about/what-is-orcid/mission.* Accessed October 1, 2017.

26. Casrai. CRediT: Contributor Roles Taxonomy. Available from: *http:// docs.casrai.org/ CRediT.* Accessed October 1, 2017.

Annexure

Parameter	Conventional unit	SI unit
Acid phosphatase	units/L	U/L
Alanine aminotransferase (ALT)	units/L	U/L
Albumin	g/dL	g/L
Alkaline phosphatase	units/L	U/L
Ammonia (as NH_3)	μg/dL	μmol/L
Amylase	units/L	U/L
Aspartate aminotransferase (AST)	units/L	U/L
Bicarbonate	mEq/L	mmol/L
Bilirubin	mg/dL	μmol/L
$PaCO_2$ mm	Hg mm	Hg
pH	pH units	pH units
PaO_2 mm	Hg mm	Hg
Calcium	mg/dL, mEq/L	mmol/L
Carbon dioxide	mEq/L	mmol/L
Ceruloplasmin	mg/dL	mg/L
Chloride	mEq/L	mmol/L
Cholesterol	mg/dL	mmol/L
Corticotropin (ACTH)	pg/mL	pmol/L
Cortisol	μg/dL	nmol/L
Creatine	mg/dL	μmol/L
Creatine kinase (CK)	units/L	U/L
Creatinine	mg/dL	μmol/L
Creatinine clearance	mL/min	mL/s
Erythrocyte sedimentation rate	mm/h	mm/h
Estradiol	pg/mL	pmol/L
Estriol	ng/mL	nmol/L
Estrone	ng/dL	pmoI/L
Ferritin	ng/mL	pmol/L
α-fetoprotein	ng/mL	μg/L
Follicle-stimulating hormone	mIU/mL	IU/L
Glucose	mg/dL	mmol/L
Hematocrit	%	proportion of 1.0
Hemoglobin (whole blood)	g/dL	g/L

Contd.

Insulin	μIU/mL	pmol/L
Iron, total	μg/dL	μmol/L
Lead	μg/dL	μmol/L
Lipids (total)	mg/dL	g/L
Lipoprotein (a)	mg/dL	μmol/L
Magnesium	mg/dL	mEq/L mmol/L
Nitrogen, nonprotein	mg/dL	mmol/L
Osmolality	mOsm/kg	mmol/kg
Parathyroid hormone	pg/mL	ng/L
Phenobarbital	mg/L	μmol/L
Phenytoin	μg/mL	μmol/L
Phosphorus	mg/dL	mmol/L
Platelets (thrombocytes)	$\times 10^3/\mu L$	$\times 10^9/L$
Potassium	mEq/L	mmol/L
Progesterone	ng/mL	nmol/L
Prolactin	μg/L	pmol
Protein, total	g/dL	g/L
Prothrombin time (PT)	s	s
Protoporphyrin, erythrocyte	μg/dL	μmol/L
Red blood cell count	$\times 10^6/\mu L$	$\times 10^{12}/L$
Reticulocyte count	% of RBCs	Proportion of 1.0
Sodium	mEq/L	mmol/L
Testosterone	ng/dL	nmol/L
Thyroglobulin	ng/mL	μg/L
TSH	mIU/L	mIU/L
Thyroxine, free (T_4)	ng/dL	pmol/L
Thyroxine, total (T_4)	μg/dL	nmol/L
Transferrin	mg/dL	g/L
Triglycerides	mg/dL	mmol/L
Triiodothyronine free (T_3)	pg/dL	pmol/L
Total (T_3)	ng/dL	nmol/L
Urea nitrogen	mg/dL	mmol/L
Uric acid	mg/dL	μmol/L
Vitamin A (retinol)	μg/dL	μmol/L
Vitamin B_6 (pyridoxine)	ng/mL	nmol/L
Vitamin B_{12} (cyanocobalamin)	pg/mL	pmol/L
Vitamin C (ascorbic acid)	mg/dL	μmol/L
Vitamin D (1,25-dihydroxyvitamin D)	pg/mL	pmol/L
Vitamin D (25-hydroxyvitamin D)	ng/mL	nmol/L
Vitamin E	mg/dL	μmol/L
Vitamin K	ng/mL	nmol/L
White blood cell count	$\times 10^3/\mu L$	$\times 10^9/L$
White blood cell differential count	%	proportion of 1.0
Zinc	μg/dL	μmol/L

Suggested Resources

1. General Dictionaries

- Acronym Finder, *http://www.acronymfinder.com*
- The American Heritage Dictionary of the English Language. 4th ed. Boston, MA: Houghton Mifflin Co; 2000.
- Dictionary.com. *http://www.dictionary.com*
- Memam-Webster's Collegiate Dictionary, llth ed. Springfield, MA: Merriam-Webster Inc; 2003. *http://www.m-w.com/dictionary.htm*
- OneLook. *http://www.onelook.com*
- Oxford Dictionaries Online, *http://www.askoxford.com*
- Your Dictionary. *com. http://www.yourdictionary.com*

2. Medical and Scientific Dictionaries

- BioTech Life Science Dictionary, *http://biotech.icmb.utexas.edu/search/dict-search.html*
- Borland's Illustrated Medical Dictionary. 30th ed. Philadelphia, PA: Saunders; 2003.
- Jablonski S. Dictionary of Medical Acronyms & Abbreviations. 5th ed. Philadelphia, PA: Hanley & Belfus Inc; 2004.
- Stedman's Medical Dictionary. 28th ed. Baltimore: Lippincott Williams & Wilkins; 2005.

3. General Style and Usage

- Acronym Finder. http://www. acronymfinder.com
- Bernstein T. The Careful Writer: A Modem Guide to English Usage. New York, NY: Free Press; 1998.
- Brooks BS, PinsonJL. Working With Words: A Concise Handbook for Media Writers and Editors. 4th ed. New York, NY: Bedford/St Martins Press; 1999.
- The Chicago Manual of Style: The Essential Guide for Writers, Editors, and Publishers. 15th ed. Chicago, IL: University of Chicago Press; 2003.
- Follett W, Wensberg E, ed. Modem American Usage: A Guide. New York, NY: Hill & Wang; 1998.
- Fowler HW, Burchfield RW, ed. The New Fowler's Modern English Usage. 3rd rev ed. New York, NY: Oxford University Press; 2000.
- Garner BA, ed. A Dictionary of Modern English Usage. 2nd ed. New York, NY: Oxford University Press; 1998.
- Garner BA. Gamer's Modern American Usage. 2nd ed. New York, NY: Oxford University Press; 2003.

- Kasdorf WE. The Columbia Guide to Digital Publishing. New York, NY: Columbia University Press; 2003.
- Lederer R, Dowis R. Sleeping Dogs Don't Lay: Practical Advice for the Grammatically Challenged, and That's No Lie. New York, NY: St Martins Press; 2001.
- McGovern G, Norton R, O'Dowd C. The Web Content Style Guide: An Essential Reference for Online Writers, Editors and Managers. London: Prentice Hall; 2001.
- Pavlicin K, Lyon C. Online Style Guide: Terms, Usage, and Tips. St Paul, MN: Elva Resa Publishing; 1998.
- Walker JR, Taylor T. The Columbia Guide to Online Style. New York, NY: Columbia University Press; 1998.
- Webster's Dictionary of English Usage. Springfield, MA: Merriam-Webster Inc; 2002.

4. Medical/Scientific Style and Usage

- American Psychological Association. Publication Manual of the American Psychological Association. 5th ed. Washington, DC: American Psychological Association; 2001.
- ASM Style Manual for Journals and Books, Washington, DC: American Society of Microbiology; 1992.
- Cohn V, Cope L. News & Numbers: A Guide to Reporting Statistical Claims and Controversies in Health and Other Fields. 2nd ed. Ames: Iowa Press; 2001.
- Coghill AM, Garson LR, eds. The ACS Style Guide: Effective Communication of Scientific Information. 3rd ed. New York, NY: Oxford University Press; 2006.
- Davis NM. Medical Abbreviations: 28,000 Conveniences at the Expense of Communications and Safety. 13th ed. Huntingdon Valley, PA: 2006.
- Rubens P, ed. Science and Technical Writing: A Manual of Style. 2nd ed. New York, NY: Routledge; 2001.
- Style Manual Committee, Council of Science Editors. Scientific Style and Format: The CSE Manual for Authors, Editors, and Publishers. 7th ed. New York, NY: Rockefeller University Press, in cooperation with the Council of Science Editors, Reston, VA; 2006.
- Sutcliffe AJ, ed. The New York Public Library Writer's Guide to Style and Usage. New York, NY: HarperCollins Publishers; 1994.

5. Writing

- Albert T. A-Z of Medical Writing. London, England: BMJ Books; 2000.
- Day RA, Gastel B. How to Write and Publish a Scientific Paper. 6th ed. Westport, CT: Greenwood Press; 2006.
- Day RA. Scientific English: A Guide for Scientists and Other Professionals. 2nd ed. Phoenix, AZ: Oryx Press; 1995.

- Gordon KE. The Deluxe Transitive Vampire: The Ultimate Handbook of Grammar for the Innocent, the Eager, and the Doomed. New York, NY: Pantheon Books; 1993.
- Gordon KE. The New Well-Tempered Sentence: A Punctuation Handbook for the Innocent, the Eager, and the Doomed. Rev ed. New Haven, CT: Ticknor & Fields; 1993.
- Huth EJ. Writing and Publishing in Medicine. 3rd ed. Baltimore, MD: Lippincott Williams & Wilkins; 1999.
- Iles RL. Guidebook to Better Medical Writing. Olathe, KS: Island Press; 1997.
- Longman Language Activator: Helps You Write and Speak Natural English. 2nd ed. White Plains, NY: Addison Wesley; 2000.
- Lunsford AA. EasyWriter: A Pocket Guide. 2nd ed. Boston, MA: Bedford/St Martins Press; 2002.
- Miller C, Swift K. Tloe Handbook of Nonsexist Writing. 2nd ed. Lincoln, NE: Universe; 2001.
- Moxley JM. Publish, Don't Perish: The Scholar's Guide to Academic Writing and Publishing. Westport, CT: Praeger Publishers; 1992.
- O'Conner P. Woe Is I: The Grammarphobe's Guide to Better English in Plain English. Expanded ed. New York, NY: Riverhead Books; 2003.
- Penrose AM, Katz SB. Writing in the Sciences: Exploring Conventions of Scientific Discourse. New York, NY: Longman; 2004.
- Strunk WJr, White EB. The Elements of Style. 4th ed. New York, NY: Allyn & Bacon; 2000.
- Truss. LEats, Shoots and Leaves: The Zero Tolerance Approach to Punctuation. New York, NY: Gotham Books; 2004.
- Wallraff B. Word Court: Wherein Verbal Virtue Is Rewarded, Crimes Against the Language Are Punished, and Poetic Justice Is Done. New York, NY: Harcourtlnc; 2000.
- Wallraff B. Your Own Words. Boulder, CO: Counterpoint Press; 2004.
- Walsh B. Lapsing Into a Comma: A Curmudgeon's Guide to the Many Things That Can Go Wrong in Print—and How to Avoid Them. Lincolnwood, IL: Contemporary Books; 2000.
- Warriner JE. English Grammar and Composition: Complete Course. Franklin ed. New York, NY: Harcourt Brace Jovanovich Publishers; 1988.
- Williams J. Style: Ten Lessons in Clarity and Grace. 6th ed. New York, NY: Longman; 2000.

6. Ethical and Legal Concerns

- Fischer MA, Perle EG, Williams JT. Perle and Williams on Publishing Law. 3rd ed. Englewood Cliffs, NJ: Aspen Law & Business; 2006.
- Garner BA. Elements of Legal Style, 2nd ed, New York, NY: Oxford University Press; 2002.
- Goldstein N, ed. Associated Press Stylebook and Briefing on Media Law. Rev and updated ed. Cambridge, MA: Perseus Publishing; 2007.

- Goldstein P. Copyright's Highway: From Gutenberg to the Celestial Jukebox. Rev ed. Stanford, CA: Stanford University Press; 2003.
- Hart JD. Law of the Web: A Field Guide to Internet Publishing. Denver, CO: Bradford Publishing Co; 2003.
- Hudson Jones A, McLellan F, (eds). Ethical Issues in Biomedical Publication. Baltimore, MD: Johns Hopkins University Press; 2000.
- International Trademark Association, http://www.inta.org/
- Maggio R. Talking About People: A Guide to Fair and Accurate Language. Phoenix, AZ: Oryx Press; 1997.
- Schwartz M; Task Force on Bias-Free Language of the Association of American University Presses. Guidelines for bias-free writing. Indiana University Press, 1995.

7. Peer Review

- Fifth International Congress on Peer Review and Biomedical Publication. *http://www.jama-peer.org*
- Godlee F, Jefferson T, (eds). Peer Review in Health Sciences. BMJ Books; 2003.
- Weller A. Editorial Peer Review: Its Strengths and Weaknesses. Medford, NJ: Information Today Inc; 2001.

8. Illustrations/Displaying Data

- Briscoe MH. Preparing Scientific Illustrations: A Guide to Better Posters, Presentations, and Publications. 2nd ed. New York, NY: Springer-Verlag; 1996.
- Cleveland WS. The Elements of Graphing Data. Summit, NJ: Hobart Press; 1994.
- Cleveland WS. Visualizing Data. Summit, NJ: Hobart Press; 1993.
- Frankel F. Envisioning Science: The Design and Craft of the Science Image. Cambridge, MA: MIT Press; 2002.
- Harris RL. Information Graphics: A Comprehensive Illustrated Reference. New York, NY: Oxford University Press; 2000.
- Tufte ER. The Cognitive Style of PowerPoint. Cheshire, CT: Graphics Press; 2003. Tufte ER. Envisioning Information. Cheshire, CT: Graphics Press; 1990.
- Tufte ER. Visual Explanations: Images and Quantities, Evidence and Narrative. Cheshire, CT: Graphics Press; 1997.

9. Databases

- Biosis: *http://www.biosis.org.* Biological/biochemical information
- CABI Publishing: *http://www. cabi-publishing.org/.* Abstracts/databases
- Centers for Disease Control and Prevention: http://www.cdc.gov
- Cochrane Library: *http://www.update-software.com/cochrane*
- EMBASE.com: *http://www. embase .com*
- Gale Directory of Online, Portable, and Internet Databases: *http:// library.dialog.com/bluesheets/html/bl0230.html*

- Human Genome Organization (HUGO): *http://www.gene.ucl.ac. uk/hugo.*
- Institute of Medicine: *http://www.nas.edu/iom*
- MEDNDX: *http: //www. medicalndx. com/* Medical search engine
- Medstract. *org: http://www.medstract.org*
- The Merck Manual: *http://www.merck.com/mrkshared/mmanual/home.jsp*
- National Academy of Sciences: *http://www.nas.edu*
- OncoLink: *http://cancer.med.upenn.edu*
- Physician's Guide to the Internet: *http://physiciansguide.com*
- ProMED-mail: *http: //www. promedmail. org*
- Thomson Scientific: *http://www.isinet.com*
- US National Library of Medicine Databases: *http://www.nlm.nih.gov*
- Includes MEDLINE, and PubMed
- World Health Organization: *http://www.who.int/en*

10. Guidelines

- CONSORT: *http://www.consort-statement.org;* A research tool that takes an evidencebased approach to improve the quality of reports of randomized trials
- The COPE Report: *http://www.publicationethics.org.uk* Committee on Publication Ethics
- Declaration of Helsinki: *http://www.wma.net/e/policy/b3.htm;* Ethical principles for medical research involving human subjects
- GPP Guidelines: *http://www.gpp-guidelines.org;* The Good Publication Practice (GPP) guidelines encourage responsible and ethical publication of the results of clinical trials sponsored by pharmaceutical companies.
- ICMJE Uniform Requirements: *http://www.ICMJE.org;* International Committee of Medical Journal Editors (*see* Appendix 2)
- PhRMA Principles: *http://www.phrma.org/clinical_trials/;* Principles on conduct of clinical trials and communication of clinical trial results
- WAME Policy Statements: *http://www.wame.org/resources/policies;* World Association of Medical Editors

11. Professional Scientific Writing, Editing, and Communications Organizations and Groups

- Committee on Publication Ethics (COPE) E-mail: *cope@bmjgroup.com;* Website: *http://www.publicationethics. org.uk*
- Council of Science Editors (CSE) E-mail: *CSE@CouncilScience Editors.org; Website: http://www.councilscienceeditors.org*
- International Committee of Medical Journal Editors (ICMJE) E-mail: *claine@acponline.org; Website. http://www.icmje.org*
- World Association of Medical Editors (WAME): *Website: http:// www.wame.org*

Appendix 1

RECOMMENDATIONS FOR THE CONDUCT, REPORTING, EDITING, AND PUBLICATION OF SCHOLARLY WORK IN MEDICAL JOURNALS

This is a reprint of the ICMJE recommendations for the conduct, reporting, editing and publication of scholarly work in medical journal. Indian pediatrics is reprinting these recommendations with permission from the ICMJE. The ICMJE has not endorsed nor approved the contents of this reprint. The official and updated version of the recommendations for the conduct, reporting, editing and publication of scholarly work in medical journal is located at *www. ICMJE.org.* Users should cite this official version when citing the document.

I. About the recommendations

a. *Purpose of the recommendations:* ICMJE developed these recommendations to review best practice and ethical standards in the conduct and reporting of research and other material published in medical journals, and to help authors, editors, and others involved in peer review and biomedical publishing create and distribute accurate, clear, reproducible, unbiased medical journal articles. The recommendations may also provide useful insights into the medical editing and publishing process for the media, patients and their families, and general readers.

b. *Who should use the recommendations:* These recommendations are intended primarily for use by authors who might submit their work for publication to ICMJE member journals. Many non-ICMJE journals voluntarily use these recommendations (see www.icmje.org/journals-following-the-icjme-recommendations/). The ICMJE encourages that use but has no authority to monitor or enforce it. In all cases, authors should use these recommendations along with individual journals' instructions to authors. Authors should also consult guidelines for the reporting of specific study types (e.g., the CONSORT guidelines for the reporting of randomized trials); see www.equator-network.org.

Journals that follow these recommendations are encouraged to incorporate them into their instructions to authors and to make explicit in those instructions that they follow ICMJE recommendations. Journals that wish to be identified on the ICMJE website as following these recommendations should notify the ICMJE secretariat at

International Committee of Medical Journal Editors (homepage on the internet). Recommendations for the conduct, reporting, editing and publication of scholarly work in medical journals. Available from: *http//www.ICMJE.org.* Accessed on 31 may, 2019.

www.icmje.org/journals-following-the-icmje-recommendations/ journal-listing-request-form/. Journals that in the past have requested such identification but who no longer follow ICMJE recommendations should use the same means to request removal from this list. The ICMJE encourages wide dissemination of these recommendations and reproduction of this document in its entirety for educational, not-for-profit purposes without regard for copyright, but all uses of the recommendations and document should direct readers to www.icmje.org for the official, most recent version, as the ICMJE updates the recommendations periodically when new issues arise.

c. *History of the recommendations:* The ICMJE has produced multiple editions of this document, previously known as the uniform requirements for manuscripts submitted to biomedical journals (URMs). The URM was first published in 1978 as a way of standardizing manuscript format and preparation across journals. Over the years, issues in publishing that went well beyond manuscript preparation arose, resulting in the development of separate statements, up-dates to the document, and its renaming as "recommendations for the conduct, reporting, editing, and publication of scholarly Work in Medical Journals" to reflect its broader scope. Previous versions of the document may be found in the "Archives" section of www.icmje.org.

II. Roles and responsibilities of authors, contributors, reviewers, editors, publishers, and owners

A. Defining the role of authors and contributors

1. *Why authorship matters*

Authorship confers credit and has important academic, social, and financial implications. Authorship also implies responsibility and accountability for published work. The following recommendations are intended to ensure that contributors who have made substantive intellectual contributions to a paper are given credit as authors, but also that contributors credited as authors understand their role in taking responsibility and being accountable for what is published. Because authorship does not communicate what contributions qualified an individual to be an author, some journals now request and publish information about the contributions of each person named as having participated in a submitted study, at least for original research. Editors are strongly encouraged to develop and implement a contributorship policy. Such policies remove much of the ambiguity surrounding contributions, but leave unresolved the question of the quantity and quality of contribution that qualify an individual for authorship. The ICMJE has thus developed criteria for authorship that can be used by all journals, including those that distinguish authors from other contributors.

2. *Who is an author?*

The ICMJE recommends that authorship be based on the following 4 criteria:

1. Substantial contributions to the conception or design of the work; or the acquisition, analysis, or interpretation of data for the work; and
2. Drafting the work or revising it critically for important intellectual content; and
3. Final approval of the version to be published; and
4. Agreement to be accountable for all aspects of the work in ensuring that questions related to the accuracy or integrity of any part of the work are appropriately investigated and resolved.

In addition to being accountable for the parts of the work he or she has done, an author should be able to identify which co-authors are responsible for specific other parts of the work. In addition, authors should have confidence in the integrity of the contributions of their co-authors.

All those designated as authors should meet all four criteria for authorship, and all who meet the four criteria should be identified as authors. Those who do not meet all four criteria should be acknowledged—see Section IIA.3 below. These authorship criteria are intended to reserve the status of authorship for those who deserve credit and can take responsibility for the work. The criteria are not intended for use as a means to disqualify colleagues from authorship who otherwise meet authorship criteria by denying them the opportunity to meet criterion #s 2 or 3. Therefore, all individuals who meet the first criterion should have the opportunity to participate in the review, drafting, and final approval of the manuscript.

The individuals who conduct the work are responsible for identifying who meets these criteria and ideally should do so when planning the work, making modifications as appropriate as the work progresses. We encourage collaboration and co-authorship with colleagues in the locations where the research is conducted. It is the collective responsibility of the authors, not the journal to which the work is submitted, to determine that all people named as authors meet all four criteria; it is not the role of journal editors to determine who qualifies or does not qualify for authorship or to arbitrate authorship conflicts. If agreement cannot be reached about who qualifies for authorship, the institution (s) where the work was performed, not the journal editor, should be asked to investigate. If authors request removal or addition of an author after manuscript submission or publication, journal editors should seek an explanation and signed statement of agreement for the requested change from all listed authors and from the author to be removed or added.

The corresponding author is the one individual who takes primary responsibility for communication with the journal during the manuscript submission, peer review, and publication process, and typically ensures that all the journal's administrative requirements, such as providing details of authorship, ethics committee approval, clinical trial registration documentation, and gathering conflict of interest forms and statements, are properly completed, although these duties may be delegated to one or more coauthors. The corresponding author should be available throughout the submission and peer-review process to respond to editorial queries in a timely way, and should be available after publication to respond to critiques of the work and cooperate with any requests from the journal for data or additional information should questions about the paper arise after publication. Although the corresponding author has primary responsibility for correspondence with the journal, the ICMJE recommends that editors send copies of all correspondence to all listed authors.

When a large multi-author group has conducted the work, the group ideally should decide who will be an author before the work is started and confirm who is an author before submitting the manuscript for publication.

All members of the group named as authors should meet all four criteria for authorship, including approval of the final manuscript, and they should be able to take public responsibility for the work and should have full confidence in the accuracy and integrity of the work of other group authors. They will also be expected as individuals to complete conflict-of-interest disclosure forms.

Some large multi-author groups designate authorship by a group name, with or without the names of individuals. When submitting a manuscript authored by a group, the corresponding author should specify the group name if one exists, and clearly identify the group members who can take credit and responsibility for the work as authors. The byline of the article identifies who is directly responsible for the manuscript, and Medline lists as authors whichever names appear on the byline. If the byline includes a group name, Medline will list the names of individual group members who are authors or who are collaborators, sometimes called non-author contributors, if there is a note associated with the byline clearly stating that the individual names are elsewhere in the paper and whether those names are authors or collaborators.

3. *Non-author contributors*
Contributors who meet fewer than all 4 of the above criteria for authorship should not be listed as authors, but they should be acknowledged. Examples of activities that alone (without other contributions) do not qualify a contributor for authorship are acquisition of funding; general supervision of a research group or general administrative support; and writing assistance, technical editing, language editing, and proofreading. Those whose contributions do not justify authorship may be acknowledged individually or together as a group under a single heading (e.g., "Clinical Investigators" or "Participating Investigators"), and their contributions should be specified (e.g., "served as scientific advisors," "critically reviewed the study proposal," "collected data," "provided and cared for study patients", "participated in writing or technical editing of the manuscript"). Because acknowledgment may imply endorsement by acknowledged individuals of a study's data and conclusions, editors are advised to require that the corresponding author obtain written permission to be acknowledged from all acknowledged individuals.

B. Conflicts of interest

Public trust in the scientific process and the credibility of published articles depend in part on how transparently conflicts of interest are handled during the planning, implementation, writing, peer review, editing, and publication of scientific work.

A conflict of interest exists when professional judgment concerning a primary interest (such as patients' welfare or the validity of research) may be influenced by a secondary interest (such as financial gain). Perceptions of conflict of interest are as important as actual conflicts of interest.

Financial relationships (such as employment, consultancies, stock ownership or options, honoraria, patents, and paid expert testimony) are the most easily identifiable conflicts of interest and the most likely to undermine the credibility of the journal, the authors, and science itself. However, conflicts can occur for other reasons, such as personal relationships or rivalries, academic competition, and intellectual beliefs. Authors should avoid entering into agreements with study sponsors, both for profit and nonprofit, that interfere with authors' access to all of the study's data or that interfere with their ability to analyze and interpret the data and to prepare and publish manuscripts independently when and where they choose. Authors may be required to provide the journal with the agreements in confidence.

Purposeful failure to disclose conflicts of interest is a form of misconduct, as is discussed in Section IIIB.

1. *Participants*

All participants in the peer-review and publication process—not only authors but also peer reviewers, editors, and editorial board members of journals— must consider their conflicts of interest when fulfilling their roles in the process of article review and publication and must disclose all relationships that could be viewed as potential conflicts of interest.

a. ***Authors:*** When authors submit a manuscript of any type or format they are responsible for disclosing all financial and recommendations for the conduct, reporting, editing, and publication of scholarly work in medical journals personal relationships that might bias or be seen to bias their work. The ICMJE has developed a form for disclosure of Conflicts of Interest to facilitate and standardize authors' disclosures. ICMJE member journals require that authors use this form, and ICMJE encourages other journals to adopt it.

b. ***Peer reviewers:*** Reviewers should be asked at the time they are asked to critique a manuscript if they have conflicts of interest that could complicate their review. Reviewers must disclose to editors any conflicts of interest that could bias their opinions of the manuscript, and should recuse themselves from reviewing specific manuscripts if the potential for bias exists. Reviewers must not use knowledge of the work they are reviewing before its publication to further their own interests.

c. ***Editors and journal staff:*** Editors who make final decisions about manuscripts should recuse themselves from editorial decisions if they have conflicts of interest or relationships that pose potential conflicts related to articles under consideration. Other editorial staff members who participate in editorial decisions must provide editors with a current description of their financial interests or other conflicts (as they might relate to editorial judgments) and recuse themselves from any decisions in which a conflict of interest exists. Editorial staff must not use information gained through working with manuscripts for private gain. Editors should publish regular disclosure statements about potential conflicts of interests related to their own commitments and those of their journal staff. Guest editors should follow these same procedures. Journals should take extra precautions and have a stated policy for evaluation of manuscripts submitted by individuals involved in editorial decisions. Further guidance is available from COPE *(https://publicationethics.org/files/A_Short_Guide_to_Ethical_ Editing. pdf) and WAME (http://wame.org/conflict-of-interest-in-peer-reviewed-medical-journals).*

2. *Reporting conflicts of interest*

Articles should be published with statements or supporting documents, such as the ICMJE conflict of interest form, declaring:

* Authors' conflicts of interest; and

* Sources of support for the work, including sponsor names along with explanations of the role of those sources if any in study design; collection, analysis, and interpretation of data; writing of the report; the decision to submit the report for publication; or a statement declaring that the supporting source had no such involvement; and

* Whether the authors had access to the study data, with an explanation of the nature and extent of access, including whether access is ongoing.

To support the above statements, editors may request that authors of a study sponsored by a funder with a proprietary or financial interest in the outcome sign a statement, such as "I had full access to all of the data in this study and I take complete responsibility for the integrity of the data and the accuracy of the data analysis."

C. **Responsibilities in the submission and peer-review process**

1. *Authors*

Authors should abide by all principles of authorship and declaration of conflicts of interest detailed in section IIA and B of this document.

a. **Predatory or pseudo-journals:** A growing number of entities are advertising themselves as "scholarly medical journals" yet do not function as such. These journals ("predatory"or "pseudo-journals") accept and publish almost all submissions and charge article processing (or publication) fees, often informing authors about this after a paper's acceptance for publication. They often claim to perform peer review but do not and may purposefully use names similar to well established journals. They may state that they are members of ICMJE but are not (see www.icmje.org for current members of the ICMJE) and that they follow the recommendations of organizations such as the ICMJE, COPE and WAME. Researchers must be aware of the existence of such entities and avoid submitting research to them for publication. Authors have a responsibility to evaluate the integrity, history, practices and reputation of the journals to which they submit manuscripts. Guidance from various organizations is available to help identify the characteristics of reputable peer-reviewed journals (www.wame.org/identifyingpredatory-or-pseudo-journals and www.wame.org/about/principlesof-transparency-and-best-practice). Seeking the assistance of scientific mentors, senior colleagues and others with many years of scholarly publishing experience may also be helpful.

2. *Journals*
a. *Confidentiality:* Manuscripts submitted to journals are privileged communications that are authors' private, confidential property, and authors may be harmed by premature disclosure of any or all of a manuscript's details. Editors therefore must not share information about manuscripts, including whether they have been received and are under review, their content and status in the review process, criticism by reviewers, and their ultimate fate, to anyone other than the authors and reviewers. Requests from third parties to use manuscripts and reviews for legal proceedings should be politely refused, and editors should recommendations for the conduct, reporting, editing, and publication of scholarly work in medical journals do their best not to provide such confidential material should it be subpoenaed.

Editors must also make clear that reviewers should keep manuscripts, associated material, and the information they contain strictly confidential. Reviewers and editorial staff members must not publicly discuss the authors' work, and reviewers must not appropriate authors' ideas before the manuscript is published. Reviewers must not retain the manuscript for their personal use and should destroy paper copies of manuscripts and delete electronic copies after submitting their reviews.

When a manuscript is rejected, it is best practice for journals to delete copies of it from their editorial systems unless retention is required by local regulations. Journals that retain copies of rejected manuscripts should disclose this practice in their Information for Authors.

When a manuscript is published, journals should keep copies of the original submission, reviews, revisions, and correspondence for at least three years and possibly in perpetuity, depending on local regulations, to help answer future questions about the work should they arise.

Editors should not publish or publicize peer reviewers' comments without permission of the reviewer and author. If journal policy is to blind authors to reviewer identity and comments are not signed, that identity must not be revealed to the author or anyone else without the reviewers' expressed written permission.

Confidentiality may have to be breached if dishonesty or fraud is alleged, but editors should notify authors or reviewers if they intend to do so and confidentiality must otherwise be honored.

b. *Timeliness:* Editors should do all they can to ensure timely processing of manuscripts with the resources available to them. If editors intend to publish a manuscript, they should attempt to do so in a timely manner and any planned delays should be negotiated with the authors. If a journal has no intention of proceeding with a manuscript, editors should endeavor to reject the manuscript as soon as possible to allow authors to submit to a different journal.

c. *Peer review:* Peer review is the critical assessment of manuscripts submitted to journals by experts who are usually not part of the editorial staff. Because unbiased, independent, critical assessment is an intrinsic part of all scholarly work, including scientific research, peer review is an important extension of the scientific process. The actual value of peer review is widely debated, but the process facilitates a fair hearing for a manuscript among members of the scientific community. More practically, it helps editors decide which manuscripts are suitable for their journals. Peer review often helps authors and editors improve the quality of reporting.

It is the responsibility of the journal to ensure that systems are in place for selection of appropriate reviewers. It is the responsibility of the editor to ensure that reviewers have access to all materials that may be relevant to the evaluation of the manuscript, including supplementary material for e-only publication, and to ensure that reviewer comments are properly assessed and interpreted in the context of their declared conflicts of interest.

A peer-reviewed journal is under no obligation to send submitted manuscripts for review, and under no obligation to follow reviewer recommendations, favorable or negative. The editor of a journal is ultimately responsible for the selection of all its content, and editorial decisions may be informed by issues unrelated to the quality of a manuscript, such as suitability for the journal. An editor can reject any article at any time before publication, including after acceptance if concerns arise about the integrity of the work. Journals may differ in the number and kinds of manuscripts they send for review, the number and types of reviewers they seek for each manuscript, whether the review process is open or blinded, and other aspects of the review process. For this reason and as a service to authors, journals should publish a description of their peer-review process. Journals should notify reviewers of the ultimate decision to accept or reject a paper, and should acknowledge the contribution of peer reviewers to their journal. Editors are encouraged to share reviewers' comments with coreviewers of the same paper, so reviewers can learn from each other in the review process.

As part of peer review, editors are encouraged to review research protocols, plans for statistical analysis if separate from the protocol, and/or contracts associated with project-specific studies. Editors should encourage authors to make such documents publicly available at the time of or after publication, before accepting such studies for publication. Some journals may require public posting of these documents as a condition of acceptance for publication.

Journal requirements for independent data analysis and for public data availability are in flux at the time of this revision, reflecting evolving views of the importance of data availability for pre- and post-publication peer review. Some journal editors currently request a statistical analysis of trial data by an independent biostatistician before accepting studies for publication. Others ask authors to say whether the study data are available to third parties to view and/or use/reanalyze, while still others encourage or require authors to share their data with others for review or reanalysis. Each journal should establish and publish their specific requirements for data analysis and post in a place that potential authors can easily access.

Some people believe that true scientific peer review begins only on the date a paper is published. In that spirit, medical journals should have a mechanism for readers to submit comments, questions, or criticisms about published articles, and authors have a responsibility to respond appropriately and cooperate with any requests from the journal for data or additional information should questions about the paper arise after publication (*see* Section III). ICMJE believes investigators have a duty to maintain the primary data and analytic procedures underpinning the published results for at least 10 years. The ICMJE encourages the preservation of these data in a data repository to ensure their longer-term availability.

d. *Integrity:* Editorial decisions should be based on the relevance of a manuscript to the journal and on the manuscript's originality, quality, and contribution to evidence about important questions. Those decisions should not be influenced by commercial interests, personal relationships or agendas, or findings that are negative or that credibly challenge accepted wisdom. In addition, authors should submit for publication or otherwise make publicly available, and editors should not exclude from consideration for publication, studies with findings that are not statistically significant or that have inconclusive findings. Such studies may provide evidence that, combined with that from other studies through meta-analysis, might still help answer important questions, and a public record of such negative or inconclusive findings may prevent unwarranted replication of effort or otherwise be valuable for other researchers considering similar work.

Journals should clearly state their appeals process and should have a system for responding to appeals and complaints.

e. *Journal metrics:* The journal impact factor is widely misused as a proxy for research and journal quality and as a measure of the importance of specific research projects or the merits of individual researchers, including their suitability for hiring, promotion, tenure, prizes, or research funding. ICMJE recommends that journals reduce the emphasis on impact factor as a single measure, but rather provide a range of article and journal metrics relevant to their readers and authors.

3. *Peer reviewers*

Manuscripts submitted to journals are privileged communications that are authors' private, confidential property, and authors may be harmed by premature disclosure of any or all of a manuscript's details.

Reviewers therefore should keep manuscripts and the information they contain strictly confidential. Reviewers must not publicly discuss authors' work and must not appropriate authors' ideas before the manuscript is published. Reviewers must not retain the manuscript for their personal use and should destroy copies of manuscripts after submitting their reviews.

Reviewers are expected to respond promptly to requests to review and to submit reviews within the time agreed. Reviewers' comments should be constructive, honest, and polite.

Reviewers should declare their conflicts of interest and recuse themselves from the peer-review process if a conflict exists.

D. Journal owners and editorial freedom

1. *Journal owners*

Owners and editors of medical journals share a common purpose, but they have different responsibilities, and sometimes those differences lead to conflicts.

It is the responsibility of medical journal owners to appoint and dismiss editors. Owners should provide editors at the time of their appointment with a contract that clearly states their rights and duties, authority, the general terms of their appointment, and mechanisms for resolving conflict. The editor's performance may be assessed using mutually agreed-upon measures, including but not necessarily limited to readership, manuscript submissions and handling times, and various journal metrics.

Owners should only dismiss editors for substantial reasons, such as scientific misconduct, disagreement with the long-term editorial direction of the journal, inadequate performance by agreed-upon performance metrics, or inappropriate behavior that is incompatible with a position of trust.

Appointments and dismissals should be based on evaluations by a panel of independent experts, rather than by a small number of executives of the owning organization. This is especially necessary in the case of dismissals because of the high value society places on freedom of speech within science and because it is often the responsibility of editors to challenge the status quo in ways that may conflict with the interests of the journal's owners.

A medical journal should explicitly state its governance and relationship to a journal owner (e.g., a sponsoring society).

2. *Editorial freedom*

The ICMJE adopts the world association of medical editors' definition of editorial freedom, which holds that editors-in-chief have full authority over the entire editorial content of their journal and the timing of publication of that content. Journal owners should not interfere in the evaluation, selection, scheduling, or editing of individual articles either directly or by creating an environment that strongly influences decisions. Editors should base editorial decisions on the validity of the work and its importance to the journal's readers, not on the commercial implications for the journal, and editors should be free to express critical but responsible views about all aspects of medicine without fear of retribution, even if these views conflict with the commercial goals of the publisher.

Editors-in-chief should also have the final say in decisions about which advertisements or sponsored content, including supplements, the journal will and will not carry, and they should have final say in use of the journal brand and in overall policy regarding commercial use of journal content.

Journals are encouraged to establish an independent editorial advisory board to help the editor establish and maintain editorial policy. Editors should seek to engage a broad and diverse array of authors, reviewers, editorial staff, editorial board members, and readers. To support editorial decisions and potentially controversial expressions of opinion, owners should ensure that appropriate insurance is obtained in the event of legal action against the editors, and should ensure that legal advice is available when necessary. If legal problems arise, the editor should inform their legal adviser and their owner and/or publisher as soon as possible. Editors should defend the confidentiality of authors and peer-reviewers (names and reviewer comments) in accordance with ICMJE policy (*see* Section IIC.2.a). Editors should take all reasonable steps to check the facts in journal commentary, including that in news sections and social media postings, and should ensure that staff working for the journal adhere to best journalistic practices including contemporaneous note-taking and seeking a response from all parties when possible before publication. Such practices in support of truth and public interest may be particularly relevant in defense against legal allegations of libel.

To secure editorial freedom in practice, the editor should have direct access to the highest level of ownership, not to a delegated manager or administrative officer. Editors and editors' organizations are obliged to support the concept of editorial freedom and to draw major transgressions of such freedom to the attention of the international medical, academic, and lay communities.

E. Protection of research participants

All investigators should ensure that the planning conduct and reporting of human research are in accordance with the Helsinki Declaration as revised in 2013 (www.wma.net/policies-post/wma-declaration-of-helsinkiethical-principles-for-medical-research-involving-humansubjects/). All authors should seek approval to conduct research from an independent local, regional, or national review body (e.g., ethics committee, institutional review board). If doubt exists whether the research was conducted in accordance with the Helsinki Declaration, the authors must explain the rationale for their approach and demonstrate that the local, regional, or national review body explicitly approved the doubtful aspects of the study. Approval by a responsible review body does not preclude editors from forming their own judgment whether the conduct of the research was appropriate.

Patients have a right to privacy that should not be violated without informed consent. Identifying information, including names, initials, or hospital numbers, should not be published in written descriptions, photographs, or pedigrees unless the information is essential for scientific purposes and the patient (or parent or guardian) gives written informed consent for publication. Informed consent for this purpose requires that an identifiable patient be shown the manuscript to be published. Authors should disclose to these patients whether any potential identifiable material might be available via the Internet as well as in print after publication. Patient consent should be written and archived with the journal, the authors, or both, as dictated by local regulations or laws. Applicable laws vary from locale to locale, and journals should establish their own policies with legal guidance. Since a journal that archives the consent will be aware of patient identity, some journals may decide that patient confidentiality is better guarded by having the author archive the consent and instead providing the journal with a written statement that attests that they have received and archived written patient consent.

Nonessential identifying details should be omitted. Informed consent should be obtained if there is any doubt that anonymity can be maintained. For example, masking the eye region in photographs of patients is inadequate protection of anonymity. If identifying characteristics are de-identified, authors should provide assurance, and editors should so note, that such changes do not distort scientific meaning.

The requirement for informed consent should be included in the journal's instructions for authors. When informed consent has been obtained, it should be indicated in the published article.

When reporting experiments on animals, authors should indicate whether institutional and national standards for the care and use of laboratory animals were followed. Further guidance on animal research ethics is available from the international association of veterinary editors' consensus author guidelines on animal ethics and welfare *(http://veteditors.org ethicsconsensusguidelines. html)*.

III. Publishing and editorial issues related to publication in medical journals

A. Corrections, retractions, republications, and version control

Honest errors are a part of science and publishing and require publication of a correction when they are detected. Corrections are needed for errors of fact. Matters of debate are best handled as letters to the editor, as print or electronic correspondence, or as posts in a journal-sponsored online forum. Updates of previous publications (e.g., an updated systematic review or clinical guideline) are considered a new publication rather than a version of a previously published article.

If a correction is needed, journals should follow these minimum standards:

- The journal should publish a correction notice as soon as possible detailing changes from and citing the original publication; the correction should be on an electronic or numbered print page that is included in an electronic or a print table of contents to ensure proper indexing.
- The journal should also post a new article version with details of the changes from the original version and the date(s) on which the changes were made.

- The journal should archive all prior versions of the article. This archive can be either directly accessible to readers or can be made available to the reader on request.
- Previous electronic versions should prominently note that there are more recent versions of the article.
- The citation should be to the most recent version. Pervasive errors can result from a coding problem or a miscalculation and may result in extensive inaccuracies throughout an article. If such errors do not change the direction or significance of the results, interpretations, and conclusions of the article, a correction should be published that follows the minimum standards noted above.

Errors serious enough to invalidate a paper's results and conclusions may require retraction. However, retraction with republication (also referred to as "replacement") can be considered in cases where honest error (e.g., a misclassification or miscalculation) leads to a major change in the direction or significance of the results, interpretations, and conclusions. If the error is judged to be unintentional, the underlying science appears valid, and the changed version of the paper survives further review and editorial scrutiny, then retraction with republication of the changed paper, with an explanation, allows full correction of the scientific literature. In such cases, it is helpful to show the extent of the changes in supplementary material or in an appendix, for complete transparency.

B. Scientific misconduct, expressions of concern, and retraction

Scientific misconduct in research and non-research publications includes but is not necessarily limited to data fabrication; data falsification, including deceptive manipulation of images; purposeful failure to disclose conflicts of interest; and plagiarism. Some people consider failure to publish the results of clinical trials and other human studies a form of scientific misconduct. While each of these practices is problematic, they are not equivalent. Each situation requires individual assessment by relevant stakeholders.

When scientific misconduct is alleged, or concerns are otherwise raised about the conduct or integrity of work described in submitted or published papers, the editor should initiate appropriate procedures detailed by such committees as the committee on publication ethics (COPE) (publicationethics. org/resources/flowcharts), consider informing the institutions and funders, and may choose to publish an expression of concern pending the outcomes of those procedures.

If the procedures involve an investigation at the authors' institution, the editor should seek to discover the outcome of that investigation; notify readers of the outcome if appropriate; and if the investigation proves scientific misconduct, publish a retraction of the article. There may be circumstances in which no misconduct is proven, but an exchange of letters to the editor could be published to highlight matters of debate to readers.

Expressions of concern and retractions should not simply be a letter to the editor. Rather, they should be prominently labelled, appear on an electronic or numbered print page that is included in an electronic or a print table of contents to ensure proper indexing, and include in their heading the title of the original article. Online, the retraction and original article should be linked in both directions

and the retracted article should be clearly labelled as retracted in all its forms (abstract, full text, PDF). Ideally, the authors of the retraction should be the same as those of the article, but if they are unwilling or unable the editor may under certain circumstances accept retractions by other responsible persons, or the editor may be the sole author of the retraction or expression of concern. The text of the retraction should explain why the article is being retracted and include a complete citation reference to that article. Retracted articles should remain in the public domain and be clearly labelled as retracted.

The validity of previous work by the author of a fraudulent paper cannot be assumed. Editors may ask the author's institution to assure them of the validity of other work published in their journals, or they may retract it. If this is not done, editors may choose to publish an announcement expressing concern that the validity of previously published work is uncertain.

The integrity of research may also be compromised by inappropriate methodology that could lead to retraction. See COPE flowcharts for further guidance on retractions and expressions of concern. *See* Section IVA3g.i for guidance about avoiding referencing retracted articles.

C. Copyright

Journals should make clear the type of copyright under which work will be published, and if the journal retains copyright, should detail the journal's position on the transfer of copyright for all types of content, including audio, video, protocols, and data sets. Medical journals may ask authors to transfer copyright to the journal. Some journals require transfer of a publication license. Some journals do not require transfer of copyright and rely on such vehicles as Creative Commons licenses. The copyright status of articles in a given journal can vary: Some content cannot be copyrighted (e.g., articles written by employees of some governments in the course of their work). Editors may waive copyright on other content, and some content may be protected under other agreements.

D. Overlapping publications

1. *Duplicate submission*

Authors should not submit the same manuscript, in the same or different languages, simultaneously to more than one journal. The rationale for this standard is the potential for disagreement when two (or more) journals claim the right to publish a manuscript that has been submitted simultaneously to more than one journal, and the possibility that two or more journals will unknowingly and unnecessarily undertake the work of peer review, edit the same manuscript, and publish the same article.

2. *Duplicate and prior publication*

Duplicate publication is publication of a paper that overlaps substantially with one already published, without clear, visible reference to the previous publication. Prior publication may include release of information in the public domain.

Readers of medical journals deserve to be able to trust that what they are reading is original unless there is a clear statement that the author and editor are intentionally republishing an article (which might be considered for historic or landmark papers, for example). The bases of this position are international copyright laws, ethical conduct, and cost-effective use of resources. Duplicate publication of original research is particularly problematic because it can result in inadvertent double-counting of data or inappropriate weighting of the results of a single study, which distorts the available evidence.

When authors submit a manuscript reporting work that has already been reported in large part in a published article or is contained in or closely related to another paper that has been submitted or accepted for publication elsewhere, the letter of submission should clearly say so and the authors should provide copies of the related material to help the editor decide how to handle the submission. (See also Section IVB).

This recommendation does not prevent a journal from considering a complete report that follows publication of a preliminary report, such as a letter to the editor, a preprint, or an abstract or poster displayed at a scientific meeting. It also does not prevent journals from considering a paper that has been presented at a scientific meeting but was not published in full, or that is being considered for publication in proceedings or similar format. Press reports of scheduled meetings are not usually regarded as breaches of this rule, but they may be if additional data tables or figures enrich such reports. Authors should also consider how dissemination of their findings outside of scientific presentations
at meetings may diminish the priority journal editors assign to their work.

Authors who choose to post their work on a preprint server should choose one that clearly identifies preprints as not peer-reviewed work and includes statements of conflicts of interest. It is the author's responsibility to inform a journal if the work has been previously posted on a preprint server. In addition, it is the author's (and not the journal editors') responsibility to ensure that preprints are amended to point readers to subsequent versions, including the final published article.

In the event of a public health emergency (as defined by public health officials), information with immediate implications for public health should be disseminated without concern that this will preclude subsequent consideration for publication in a journal. We encourage editors to give priority to authors who have made crucial data publicly available (e.g., in a gene bank) without delay.

Sharing with public media, government agencies, or manufacturers the scientific information described in a paper or a letter to the editor that has been accepted but not yet published violates the policies of many journals. Such reporting may be warranted when the paper or letter describes major therapeutic advances; reportable diseases; or public health hazards, such as serious adverse effects of drugs, vaccines, other biological products, medical devices. This reporting, whether in print or online, should not jeopardize publication, but should be discussed with and agreed upon by the editor in advance when possible.

The ICMJE will not consider as prior publication the posting of trial results in any registry that meets the criteria noted in Section III.L. if results are limited to a brief (500 word) structured abstract or tables (to include participants enrolled, key outcomes, and adverse events). The ICMJE encourages authors to include a statement with the registration that indicates that the results have not yet been published in a peer-reviewed journal, and to update the results registry with the full journal citation when the results are published.

Editors of different journals may together decide to simultaneously or jointly publish an article if they believe that doing so would be in the best interest of public health. However, the National Library of Medicine (NLM) indexes all such simultaneously published joint publications separately, so editors should include a statement making the simultaneous publication clear to readers.

Authors who attempt duplicate publication without such notification should expect at least prompt rejection of the submitted manuscript. If the editor was not aware of the violations and the article has already been published, then the article might warrant retraction with or without the author's explanation or approval. See COPE flowcharts for further guidance on handling duplicate publication.

3. *Acceptable secondary publication*

Secondary publication of material published in other journals or online may be justifiable and beneficial, especially when intended to disseminate important information to the widest possible audience (e.g., guidelines produced by government agencies and professional organizations in the same or a different language). Secondary publication for various other reasons may also be justifiable provided the following conditions are met:

1. The authors have received approval from the editors of both journals (the editor concerned with secondary publication must have access to the primary version).

2. The priority of the primary publication is respected by a publication interval negotiated by both editors with the authors.

3. The paper for secondary publication is intended for a different group of readers; an abbreviated version could be sufficient.

4. The secondary version faithfully reflects the data and interpretations of the primary version.

5. The secondary version informs readers, peers, and documenting agencies that the paper has been published in whole or in part elsewhere—for example, with a note that might read, "This article is based on a study first reported in the [journal title, with full reference]"—and the secondary version cites the primary reference.

6. The title of the secondary publication should indicate that it is a secondary publication (complete or abridged republication or translation) of a primary publication. Of note, the NLM does not consider translations to be

"republications" and does not cite or index them when the original article was published in a journal that is indexed in Medline.

When the same journal simultaneously publishes an article in multiple languages, the Medline citation will note the multiple languages (e.g., Angelo M. Journal networking in nursing: a challenge to be shared. Rev Esc Enferm USP. 2011 Dec 45[6]:1281-2,1279-80,1283-4. Article in English, Portuguese, and Spanish. No abstract available. PMID 22241182).

4. *Manuscripts based on the same database*

If editors receive manuscripts from separate research groups or from the same group analyzing the same data set (e.g., from a public database, or systematic reviews or meta-analyses of the same evidence), the manuscripts should be considered independently because they may differ in their analytic methods, conclusions, or both. If the data interpretation and conclusions are similar, it may be reasonable although not mandatory for editors to give preference to the manuscript submitted first. Editors might consider publishing more than one manuscript that overlap in this way because different analytical approaches may be complementary and equally valid, but manuscripts based upon the same dataset should add substantially to each other to warrant consideration for publication as separate papers, with appropriate citation of previous publications from the same dataset to allow for transparency.

Secondary analyses of clinical trial data should cite any primary publication, clearly state that it contains secondary analyses/results, and use the same identifying trial registration number as the primary trial and unique, persistent dataset identifier.

Sometimes for large trials it is planned from the beginning to produce numerous separate publications regarding separate research questions but using the same original participant sample. In this case authors may use the original single trial registration number, if all the outcome parameters were defined in the original registration. If the authors registered several substudies as separate entries in, for example, clinicaltrials.gov, then the unique trial identifier should be given for the study in question, The main issue is transparency, so no matter what model is used it should be obvious for the reader.

E. Correspondence

Medical journals should provide readers with a mechanism for submitting comments, questions, or criticisms about published articles, usually but not necessarily always through a correspondence section or online forum. The authors of articles discussed in correspondence or an online forum have a responsibility to respond to substantial criticisms of their work using those same mechanisms and should be asked by editors to respond. Authors of correspondence should be asked to declare any competing or conflicting interests.

Correspondence may be edited for length, grammatical correctness, and journal style. Alternatively, editors may choose to make available to readers unedited correspondence, for example, via an online commenting system.

Such commenting is not indexed in Medline unless, it is subsequently published on a numbered electronic or print page. However, the journal handles correspondence, it should make known its practice. In all instances, editors must make an effort to screen discourteous, inaccurate, or libellous comments.

Responsible debate, critique, and disagreement are important features of science, and journal editors should encourage such discourse ideally within their own journals about the material they have published. Editors, however, have the prerogative to reject correspondence that is irrelevant, uninteresting, or lacking cogency, but they also have a responsibility to allow a range of opinions to be expressed and to promote debate. In the interests of fairness and to keep correspondence within manageable proportions, journals may want to set time limits for responding to published material and for debate on a given topic.

F. Fees

Journals should be transparent about their types of revenue streams. Any fees or charges that are required for manuscript processing and/or publishing materials in the journal shall be clearly stated in a place that is easy for potential authors to find prior to submitting their manuscripts for review or explained to authors before they begin preparing their manuscript for submission (http://publica tionethics.org/files/u7140/Principles_of_Transparency_and_Best_Practice_in_Scholarly_Publishing.pdf).

G. Supplements, theme issues, and special series

Supplements are collections of papers that deal with related issues or topics, are published as a separate issue of the journal or as part of a regular issue, and may be funded by sources other than the journal's publisher. Because funding sources can bias the content of supplements through the choice of topics and viewpoints, journals should adopt the following principles, which also apply to theme issues or special series that have external funding and/or guest editors:

1. The journal editor must be given and must take full responsibility for the policies, practices, and content of supplements, including complete control of the decision to select authors, peer reviewers, and content for the supplement. Editing by the funding organization should not be permitted.

2. The journal editor has the right to appoint one or more external editors of the supplement and must take responsibility for the work of those editors.

3. The journal editor must retain the authority to send supplement manuscripts for external peer review and to reject manuscripts submitted for the supplement with or without external review. These conditions should be made known to authors and any external editors of the supplement before beginning editorial work on it.

4. The source of the idea for the supplement, sources of funding for the supplement's research and publication, and products of the funding source related to content considered in the supplement should be clearly stated in the introductory material.

5. Advertising in supplements should follow the same policies as those of the primary journal.
6. Journal editors must enable readers to distinguish readily between ordinary editorial pages and supplement pages.
7. Journal and supplement editors must not accept personal favors or direct remuneration from sponsors of supplements.
8. Secondary publication in supplements (republication of papers published elsewhere) should be clearly identified by the citation of the original paper and by the title.
9. The same principles of authorship and disclosure of potential conflicts of interest discussed elsewhere in this document should be applied to supplements.

H. Sponsorship or partnership
Various entities may seek interactions with journals or editors in the form of sponsorships, partnerships, meetings, or other types of activities. To preserve editorial independence, these interactions should be governed by the same principles outlined above for Supplements, Theme Issues, and Special Series (*see* Section IIIG).

I. Electronic publishing
Most medical journals are now published in electronic as well as print versions, and some are published only in electronic form. Principles of print and electronic publishing are identical, and the recommendations of this document apply equally to both. However, electronic publishing provides opportunities for versioning and raises issues about link stability and content preservation that are addressed here.

Recommendations for corrections and versioning are detailed in Section III.A.

Electronic publishing allows linking to sites and resources beyond journals over which journal editors have no editorial control. For this reason, and because links to external sites could be perceived as implying endorsement of those sites, journals should be cautious about external linking. When a journal does link to an external site, it should state that it does not endorse or take responsibility or liability for any content, advertising, products, or other materials on the linked sites, and does not take responsibility for the sites' availability.

Permanent preservation of journal articles on a journal's website, or in an independent archive or a credible repository, is essential for the historical record. Removing an article from a journal's website in its entirety is almost never justified as copies of the article may have been downloaded even if its online posting was brief. Such archives should be freely accessible or accessible to archive members. Deposition in multiple archives is encouraged. However, if necessary for legal reasons (e.g., libel action), the URL for the removed article must contain a detailed reason for the removal, and the article must be retained in the journal's internal archive.

Permanent preservation of a journal's total content is the responsibility of the journal publisher, who in the event of journal termination should be certain the journal files are transferred to a responsible third party who can make the content available.

Journal websites should post the date that nonarticle web pages, such as those listing journal staff, editorial board members, and instructions for authors, were last updated.

J. Advertising

Most medical journals carry advertising, which generates income for their publishers, but journals should not be dominated by advertisements, and advertising must not be allowed to influence editorial decisions.

Journals should have formal, explicit, written policies for advertising in both print and electronic versions. Best practice prohibits selling advertisements intended to be juxtaposed with editorial content on the same product. Advertisements should be clearly identifiable as advertisements. Editors should have full and final authority for approving print and online advertisements and for enforcing advertising policy.

Journals should not carry advertisements for products proven to be seriously harmful to health. Editors should ensure that existing regulatory or industry standards for advertisements specific to their country are enforced, or develop their own standards. The interests of organizations or agencies should not control classified and other nondisplay advertising, except where required by law. Editors should consider all criticisms of advertisements for publication.

K. Journals and the media

Journals' interactions with media should balance competing priorities. The general public has a legitimate interest in all journal content and is entitled to important information within a reasonable amount of time, and editors have a responsibility to facilitate that. However media reports of scientific research before it has been peer-reviewed and fully vetted may lead to dissemination of inaccurate or premature conclusions, and doctors in practice need to have research reports available in full detail before they can advise patients about the reports' conclusions.

An embargo system has been established in some countries and by some journals to assist this balance, and to prevent publication of stories in the general media before publication of the original research in the journal. For the media, the embargo creates a "level playing field," which most reporters and writers appreciate since it minimizes the pressure on them to publish stories before competitors when they have not had time to prepare carefully. Consistency in the timing of public release of biomedical information is also important in minimizing economic chaos, since some articles contain information that has potential to influence financial markets. The ICMJE acknowledges criticisms of embargo systems as being selfserving of journals' interests and an impediment to rapid dissemination of scientific information, but believe the benefits of the systems outweigh their harms.

The following principles apply equally to print and electronic publishing and may be useful to editors as they seek to establish policies on interactions with the media:

- Editors can foster the orderly transmission of medical information from researchers, through peer-reviewed journals, to the public. This can be accomplished by an agreement with authors that they will not publicize their work while their manuscript is under consideration or awaiting publication and an agreement with the media that they will not release stories before publication of the original research in the journal, in return for which the journal will cooperate with them in preparing accurate stories by issuing, for example, a press release.
- Editors need to keep in mind that an embargo system works on the honor system—no formal enforcement or policing mechanism exists. The decision of a significant number of media outlets or biomedical journals not to respect the embargo system would lead to its rapid dissolution.
- Notwithstanding authors' belief in their work, very little medical research has such clear and urgently important clinical implications for the public's health that the news must be released before full publication in a journal. When such exceptional circumstances occur, the appropriate authorities responsible for public health should decide whether to disseminate information to physicians and the media in advance and should be responsible for this decision. If the author and the appropriate authorities wish to have a manuscript considered by a particular journal, the editor should be consulted before any public release. If editors acknowledge the need for immediate release, they should waive their policies limiting prepublication publicity.
- Policies designed to limit prepublication publicity should not apply to accounts in the media of presentations at scientific meetings or to the abstracts from these meetings (see Duplicate Publication). Researchers who present their work at a scientific meeting should feel free to discuss their presentations with reporters but should be discouraged from offering more detail about their study than was presented in the talk, or should consider how giving such detail might diminish the priority journal editors assign to their work (see Duplicate Publication).
- When an article is close to being published, editors or journal staff should help the media prepare accurate reports by providing news releases, answering questions, supplying advance copies of the article, or referring reporters to appropriate experts. This assistance should be contingent on the media's cooperation in timing the release of a story to coincide with publication of the article.

L. Clinical trials

1. *Registration*

The ICMJE's clinical trial registration policy is detailed in a series of editorials (see Updates and Editorials [www.icmje.org/news-and-editorials/] and FAQs [www.icmje.org/abouticmje/faqs/]).

Briefly, the ICMJE requires, and recommends that all medical journal editors require, registration of clinical trials in a public trials registry at or before the time of first patient enrollment as a condition of consideration for publication. Editors requesting inclusion of their journal on the ICMJE website list of publications that follow ICMJE guidance [icmje.org/journals.html] should recognize that the listing implies enforcement by the journal of ICMJE's trial registration policy.

ICMJE uses the date trial registration materials were first submitted to a registry as the date of registration. When there is a substantial delay between the submission of registration materials and their posting at the trial registry, editors may inquire about the circumstances that led to the delay.

The ICMJE defines a clinical trial as any research project that prospectively assigns people or a group of people to an intervention, with or without concurrent comparison or control groups, to study the relationship between a health-related intervention and a health outcome. Health-related interventions are those used to modify a biomedical or health-related outcome; examples include drugs, surgical procedures, devices, behavioral treatments, educational programs, dietary interventions, quality improvement interventions, and process-of-care changes. Health outcomes are any biomedical or health-related measures obtained in patients or participants, including pharmacokinetic measures and adverse events. The ICMJE does not define the timing of first participant enrollment, but best practice dictates registration by the time of first participant consent.

The ICMJE accepts publicly accessible registration in any registry that is a primary register of the WHO International Clinical Trials Registry Platform (ICTRP) (www.who.int/ictrp/network/primary/en/index.html) or in ClinicalTrials.gov, which is a data provider to the WHO ICTRP. The ICMJE endorses these registries because they meet several criteria. They are accessible to the public at no charge, open to all prospective registrants, managed by a not-for-profit organization, have a mechanism to ensure the validity of the registration data, and are electronically searchable. An acceptable registry must include the minimum 21 item trial registration dataset (http://prsinfo.clinicaltrials.gov/train Trainer/WHO-ICMJE-ClinTrialsgov-Cross-Ref.pdf or www .who.int/ictrp/network/trds/en/index.html) at the time of registration and before enrollment of the first participant. The ICMJE considers inadequate trial registrations missing any of the 21 data fields, those that have fields that contain uninformative information, or registrations that are not made publicly accessible such as phase I trials submitted to the EU-CTR and trials of devices for which the information is placed in a "lock box." In order to comply with ICMJE policy, investigators registering trials of devices at ClinicalTrials.gov must "opt out" of the lock box by electing public posting prior to device approval. Although not a required item, the ICMJE encourages authors to include a statement that indicates that the results have not yet been published in a peer-reviewed journal, and to update the registration with the full journal citation when the results are published.

The purpose of clinical trial registration is to prevent selective publication and selective reporting of research outcomes, to prevent unnecessary duplication of research effort, to help patients and the public know what trials are planned or ongoing into which they might want to enroll, and to help give ethics review boards considering approval of new studies a view of similar work and data relevant to the research they are considering. Retrospective registration, for example at the time of manuscript submission, meets none of these purposes. Those purposes apply also to research with alternative designs, for example, observational studies. For that reason, the ICMJE encourages registration of research with non-trial designs, but because the exposure or intervention in non-trial research is not dictated by the researchers, the ICMJE does not require it.

Secondary data analyses of primary (parent) clinical trials should not be registered as separate clinical trials, but instead should reference the trial registration number of the primary trial.

The ICMJE expects authors to ensure that they have met the requirements of their funding and regulatory agencies regarding aggregate clinical trial results reporting in clinical trial registries. It is the authors', and not the journal editors', responsibility to explain any discrepancies between results reported in registries and journal publications. The ICMJE will not consider as prior publication the posting of trial results in any registry that meets the above criteria if results are limited to a brief (500 word) structured abstract or tables (to include trial participants enrolled, baseline characteristics, primary and secondary outcomes, and adverse events).

The ICMJE recommends that journals publish the trial registration number at the end of the abstract. The ICMJE also recommends that, whenever a registration number is available, authors list this number the first time they use a trial acronym to refer either to the trial they are reporting or to other trials that they mention in the manuscript.

Editors may consider whether the circumstances involved in a failure to appropriately register a clinical trial were likely to have been intended to or resulted in biased reporting. Because of the importance of prospective trial registration, if an exception to this policy is made, trials must be registered and the authors should indicate in the publication when registration was completed and why it was delayed. Editors should publish a statement indicating why an exception was allowed. The ICMJE emphasizes that such exceptions should be rare, and that authors failing to prospectively register a trial risk its inadmissibililty to our journals.

2. *Data sharing*

The ICMJE's data sharing statement policy is detailed in an editorial (see Updates and Editorials [www.icmje.org/update.html]).

1. As of 1 July 2018 manuscripts submitted to ICMJE journals that report the results of clinical trials must contain a data sharing statement as described below.

2. Clinical trials that begin enrolling participants on or after 1 January 2019 must include a data sharing plan in the trial's registration. The ICMJE's policy regarding trial registration is explained at www.icmje.org/recommendations/browse/publishing-and-editorial-issues/clinical-trial-registration.html. If the data sharing plan changes after registration this should be reflected in the statement

submitted and published with the manuscript, and updated in the registry record.

Data sharing statements must indicate the following: whether individual deidentified participant data (including data acceptable answer); what data in particular will be shared; whether additional, related documents will be available (e.g., study protocol, statistical analysis plan, etc.); when the data will become available and for how long; by what access criteria data will be shared (including with whom, for what types of analyses, and by what mechanism). Illustrative examples of data sharing statements that would meet these requirements are provided in the Table.

Authors of secondary analyses using shared data must attest that their use was in accordance with the terms (if any) agreed to upon their receipt. They must also reference the source of the data using its unique, persistent identifier to provide appropriate credit to those who generated it and allow searching for the studies it has supported. Authors of secondary analyses must explain completely how theirs differ from previous analyses. In addition, those who generate and then share clinical trial data sets deserve substantial credit for their efforts. Those using data collected by others should seek collaboration with those who collected the data. As collaboration will not always be possible, practical, or desired, the efforts of those who generated the data must be recognized.

IV. Manuscript preparation and submission

A. Preparing a manuscript for submission to a medical journal

1. *General principles*

The text of articles reporting original research is usually divided into introduction, methods, results, and discussion sections. This so-called "IMRAD" structure is not an arbitrary publication format but a reflection of the process of scientific discovery. Articles often need subheadings within these sections to further organize their content. Other types of articles, such as meta-analyses, may require different formats, while case reports, narrative reviews, and editorials may have less structured or unstructured formats.

Electronic formats have created opportunities for adding details or sections, layering information, cross-linking, or extracting portions of articles in electronic versions. Supplementary electronic-only material should be submitted and sent for peer review simultaneously with the primary manuscript.

2. *Reporting guidelines*

Reporting guidelines have been developed for different study designs; examples include CONSORT (www.consort-statement. org) for randomized trials, STROBE for observational studies (http://strobe-statement.org/), PRISMA for systematic reviews and meta-analyses (http://prisma-statement.org/), and STARD for studies of diagnostic accuracy (www.stard-statement.org/). Journals are encouraged to ask authors to follow these guidelines because they help authors describe the study in enough detail for

Table: Examples of data sharing statements that fulfill these ICMJE requirements*

	Example 1	Example 2	Example 3	Example 4
Will individual participant data be available (including data dictionaries)?	Yes	Yes	Yes	No
What data in particular will be shared?	All of the individual participant data collected during the trial, after deidentification.	Individual participant data that underlie the results reported in this article, after deidentification (text, tables, figure, and appendices).	Individual participant data that underlie the results reported in this article, after deidentification (text, tables, figures, and appendices).	Not available
What other documents will be available?	Study protocol, statistical analysis plan, informed consent form, clinical study report, analytic code	Study protocol, statistical analysis plan, analytic code	Study protocol	Not available
When will data be available (start and end dates)?	Immediately following publication. No end date.	Beginning 3 months and ending 5 years following article publication.	Beginning 9 months and ending 36 months following article publication.	Not available
With whom?	Anyone who wishes to access the data.	Researchers who provide a methodologically sound proposal.	Investigators whose proposed use of the data has been approved by an independent review committee (learned intermediary) identified for this purpose.	Not applicable

Contd.

Table: Examples of data sharing statements that fulfill these ICMJE requirements* (*Contd.*)

	Example 1	Example 2	Example 3	Example 4
For what types of analyses?	Any purpose.	To achieve aims in the approved proposal.	For individual participant data meta-analysis.	Not applicable
By what mechanism will data be made available?	Data are available indefinitely at (link to be included).	Proposals should be directed to xxx@yyy. To gain access, data requestors will need to sign a data access agreement. Data are available for 5 years at a third party website (Link to be included).	Proposals may be submitted up to 36 months following article publication. After 36 months the data will be available in our University's data warehouse but without investigator support other than deposited metadata. Information regarding submitting proposals and accessing data may be found at (Link to be provided).	Not applicable

* These examples are meant to illustrate a range of, but not all, data sharing options.

it to be evaluated by editors, reviewers, readers, and other researchers evaluating the medical literature. Authors of review manuscripts are encouraged to describe the methods used for locating, selecting, extracting, and synthesizing data; this is mandatory for systematic reviews. Good sources for reporting guidelines are the EQUATOR Network *(www.equator-network.org/home/)* and the NLM's Research Reporting Guidelines and Initiatives *(www.nlm .nih.gov/services/research_report_guide.html).*

3. *Manuscript sections*

The following are general requirements for reporting within sections of all study designs and manuscript formats.

a. *Title page:* General information about an article and its authors is presented on a manuscript title page and usually includes the article title, author information, any disclaimers, sources of support, word count, and sometimes the number of tables and figures. Article title. The title provides a distilled description of the complete article and should include information that, along with the abstract, will make electronic retrieval of the article sensitive and specific. Reporting guidelines recommend and some journals require that information about the study design be a part of the title (particularly important for randomized trials and systematic reviews and meta-analyses). Some journals require a short title, usually no more than 40 characters (including letters and spaces) on the title page or as a separate entry in an electronic submission system. Electronic submission systems may restrict the number of characters in the title.

Author information: Each author's highest academic degrees should be listed, although some journals do not publish these. The name of the department(s) and institution (s) or organizations where the work should be attributed should be specified. Most electronic submission systems require that authors provide full contact information, including land mail and e-mail addresses, but the title page should list the corresponding authors' telephone and fax numbers and e-mail address. ICMJE encourages the listing of authors' Open Researcher and Contributor Identification (ORCID).

Disclaimers: An example of a disclaimer is an author's statement that the views expressed in the submitted article are his or her own and not an official position of the institution or funder. Source(s) of support. These include grants, equipment, drugs, and/or other support that facilitated conduct of the work described in the article or the writing of the article itself.

Word count: A word count for the paper's text, excluding its abstract, acknowledgments, tables, figure legends, and references, allows editors and reviewers to assess whether the information contained in the paper warrants the paper's length, and whether the submitted manuscript fits within the journal's formats and word limits. A separate word count for the abstract is useful for the same reason.

Number of figures and tables: Some submission systems require specification of the number of figures and tables before uploading the relevant files. These numbers allow editorial staff and reviewers to confirm that all figures and tables were actually included with the manuscript and, because tables and figures occupy space, to assess if the information provided by the figures and tables warrants the paper's length and if the manuscript fits within the journal's space limits.

Conflict of interest declaration. Conflict of interest information for each author needs to be part of the manuscript; each journal should develop standards with regard to the form the information should take and where it will be posted. The ICMJE has developed a uniform conflict of interest disclosure form for use by ICMJE member journals (www.icmje.org/coi_disclosure. pdf), and the ICMJE encourages other journals to adopt it. Despite availability of the form, editors may require conflict of interest declarations on the manuscript title page to save the work of collecting forms from each author prior to making an editorial decision or to save reviewers and readers the work of reading each author's form.

b. *Abstract:* Original research, systematic reviews, and metaanalyses require structured abstracts. The abstract should provide the context or background for the study and should state the study's purpose, basic procedures (selection of study participants, settings, measurements, analytical methods), main findings (giving specific effect sizes and their statistical and clinical significance, if possible), and principal conclusions. It should emphasize new and important aspects of the study or observations, note important limitations, and not overinterpret findings. clinical trial abstracts should include items that the CONSORT group has identified as essential (www.consort-statement.org/resources/downloads/extensions/consort-extension-for-abstracts-2008pdf/). Funding sources should be listed separately after the abstract to facilitate proper display and indexing for search retrieval by MEDLINE.

Because abstracts are the only substantive portion of the article indexed in many electronic databases, and the only portion many readers read, authors need to ensure that they accurately reflect the content of the article. Unfortunately, information in abstracts often differs from that in the text. Authors and editors should work in the process of revision and review to ensure that information is consistent in both places. The format required for structured abstracts differs from journal to journal, and some journals use more than one format; authors need to prepare their abstracts in the format specified by the journal they have chosen.

The ICMJE recommends that journals publish the clinical trial registration number at the end of the abstract. The ICMJE also recommends that, when a registration number is available, authors list that number the first time they use a trial acronym to refer to the trial they are reporting or to other trials that they mention in the manuscript. If the data have been deposited in a public repository and/or are being used in a secondary analysis, authors should state at the end of the abstract the unique, persistent data set identifier; repository name; and number.

c. *Introduction:* Provide a context or background for the study (that is, the nature of the problem and its significance). State the specific purpose or research objective of, or hypothesis tested by, the study or observation. Cite only directly pertinent references, and do not include data or conclusions from the work being reported.

d. *Methods:* The guiding principle of the Methods section should be clarity about how and why a study was done in a particular way. The Methods section should aim to be sufficiently detailed such that others with access to the data would be able to reproduce the results. In general, the section should include only information that was available at the time the plan or protocol for the study was being written; all information obtained during the study belongs in the Results section. If an organization was paid or otherwise contracted to help conduct the research (examples include data collection and management), then this should be detailed in the methods. The Methods section should include a statement indicating that the research was approved by an independent local, regional or national review body (e.g., ethics committee, institutional review board). If doubt exists whether the research was conducted in accordance with the Helsinki Declaration, the authors must explain the rationale for their approach and demonstrate that the local, regional or national review body explicitly approved the doubtful aspects of the study. (*See* Section IIE).

   i. *Selection and description of participants:* Clearly describe the selection of observational or experimental participants (healthy individuals or patients, including controls), including eligibility and exclusion criteria and a description of the source population. Because the relevance of such variables as age, sex, or ethnicity is not always known at the time of study design, researchers should aim for inclusion of representative populations into all study types and at a minimum provide descriptive data for these and other relevant demographic variables. Ensure correct use of the terms sex (when reporting biological factors) and gender (identity, psychosocial or cultural factors), and, unless inappropriate, report the sex and/or gender of study participants, the sex of animals or cells, and describe the methods used to determine sex and gender. If the study was done involving an exclusive population, for example in only one sex, authors should justify why, except in obvious cases (e.g., prostate cancer). Authors should define how they determined race or ethnicity and justify their relevance. Authors should use neutral, precise, and respectful language to describe study participants and avoid the use of terminology that might stigmatize participants.

   ii. *Technical information:* Specify the study's main and secondary objectives—usually identified as primary and secondary outcomes. Identify methods, equipment (give the manufacturer's name and address in parentheses), and procedures in sufficient detail to allow others to reproduce the results. Give references to established methods, including statistical methods (see below); provide references and brief descriptions for methods that have been published but are not well-known; describe new or substantially modified methods, give the reasons for using them, and evaluate their limitations. Identify precisely all drugs and chemicals used, including generic name(s), dose(s), and route(s) of administration. Identify appropriate scientific names and gene names.

iii. *Statistics:* Describe statistical methods with enough detail to enable a knowledgeable reader with access to the original data to judge its appropriateness for the study and to verify the reported results. When possible, quantify findings and present them with appropriate indicators of measurement error or uncertainty (such as confidence intervals). Avoid relying solely on statistical hypothesis testing, such as P values, which fail to convey important information about effect size and precision of estimates. References for the design of the study and statistical methods should be to standard works when possible (with pages stated). Define statistical terms, abbreviations, and most symbols. Specify the statistical software package(s) and versions used. Distinguish prespecified from exploratory analyses, including subgroup analyses.

e. *Results:* Present your results in logical sequence in the text, tables, and figures, giving the main or most important findings first. Do not repeat all the data in the tables or figures in the text; emphasize or summarize only the most important observations. Provide data on all primary and secondary outcomes identified in the Methods section. Extra or supplementary materials and technical details can be placed in an appendix where they will be accessible but will not interrupt the flow of the text, or they can be published solely in the electronic version of the journal.

Give numeric results not only as derivatives (e.g., percentages) but also as the absolute numbers from which the derivatives were calculated, and specify the statistical significance attached to them, if any. Restrict tables and figures to those needed to explain the argument of the paper and to assess supporting data. Use graphs as an alternative to tables with many entries; do not duplicate data in graphs and tables. Avoid nontechnical uses of technical terms in statistics, such as "random" (which implies a randomizing device), "normal," "significant," "correlations," and "sample."

Separate reporting of data by demographic variables, such as age and sex, facilitate pooling of data for subgroups across studies and should be routine, unless there are compelling reasons not to stratify reporting, which should be explained.

f. *Discussion:* It is useful to begin the discussion by briefly summarizing the main findings, and explore possible mechanisms or explanations for these findings. Emphasize the new and important aspects of your study and put your findings in the context of the totality of the relevant evidence. State the limitations of your study, and explore the implications of your findings for future research and for clinical practice or policy. Discuss the influence or association of variables, such as sex and/or gender, on your findings, where appropriate, and the limitations of the data. Do not repeat in detail data or other information given in other parts of the manuscript, such as in the Introduction or the results section. Link the conclusions with the goals of the study but avoid unqualified statements and conclusions not adequately supported by the data. In particular, distinguish between clinical and statistical significance, and avoid making statements on economic benefits and costs unless the manuscript includes the appropriate economic data and analyses. Avoid claiming priority or alluding to work that has not been completed. State new hypotheses when warranted, but label them clearly.

g. *References*

　i. *General considerations:* Authors should provide direct references to original research sources whenever possible. References should not be used by authors, editors, or peer reviewers to promote self-interests. Although references to review articles can be an efficient way to guide readers to a body of literature, review articles do not always reflect original work accurately. On the other hand, extensive lists of references to original work on a topic can use excessive space. Fewer references to key original papers often serve as well as more exhaustive lists, particularly since references can now be added to the electronic version of published papers, and since electronic literature searching allows readers to retrieve published literature efficiently.

　Do not use conference abstracts as references: they can be cited in the text, in parentheses, but not as page footnotes. References to papers accepted but not yet published should be designated as "in press" or "forthcoming." Information from manuscripts submitted but not accepted should be cited in the text as "unpublished observations" with written permission from the source.

　Published articles should reference the unique, persistent identifiers of the datasets employed.

　Avoid citing a "personal communication" unless it provides essential information not available from a public source, in which case the name of the person and date of communication should be cited in parentheses in the text. For scientific articles, obtain written permission and confirmation of accuracy from the source of a personal communication.

　Some but not all journals check the accuracy of all reference citations; thus, citation errors sometimes appear in the published version of articles. To minimize such errors, references should be verified using either an electronic bibliographic source, such as PubMed, or print copies from original sources. Authors are responsible for checking that none of the references cite retracted articles except in the context of referring to the retraction. For articles published in journals indexed in Medline, the ICMJE considers PubMed the authoritative source for information about retractions. Authors can identify retracted articles in Medline by searching PubMed for "Retracted publication [pt]", where the term "pt" in square brackets stands for publication type, or by going directly to the PubMed's list of retracted publications (www.ncbi.nlm.nih.gov/pubmed?term+retracted+ publication+[pt]).

References should be numbered consecutively in the order in which they are first mentioned in the text. Identify references in text, tables, and legends by Arabic numerals in parentheses.

References cited only in tables or figure legends should be numbered in accordance with the sequence established by the first identification in the text of the particular table or figure. The titles of journals should be abbreviated according to the style used for Medline (www.ncbi.nlm.nih.gov/nlmcatalog/journals). Journals vary on whether they ask authors to cite electronic references within parentheses in the text or in numbered references following the text. Authors should consult with the journal to which they plan to submit their work.

ii. *Style and format:* References should follow the standards summarized in the NLM's international committee of medical journal editors (ICMJE) recommendations for the conduct, reporting, editing, and publication of scholarly work in medical journals: Sample References (www.nlm.nih.gov/bsd/uniform_requirements.html) webpage and detailed in the NLM's Citing Medicine, 2nd edition (www.ncbi.nlm.nih.gov/books/NBK7256/). These resources are regularly updated as new media develop, and currently include guidance for print documents; unpublished material; audio and visual media; material on CD-ROM, DVD, or disk; and material on the Internet.

h. *Tables:* Tables capture information concisely and display it efficiently; they also provide information at any desired level of detail and precision. Including data in tables rather than text frequently makes it possible to reduce the length of the text. Prepare tables according to the specific journal's requirements; to avoid errors it is best if tables can be directly imported into the journal's publication software. Number tables consecutively in the order of their first citation in the text and supply a title for each. Titles in tables should be short but self-explanatory, containing information that allows readers to understand the table's content without having to go back to the text. Be sure that each table is cited in the text.

Give each column a short or an abbreviated heading. Authors should place explanatory matter in footnotes, not in the heading. Explain all nonstandard abbreviations in footnotes, and use symbols to explain information if needed. Symbols may vary from journal to journal (alphabet letter or such symbols as *, †, ‡, §), so check each journal's instructions for authors for required practice. Identify statistical measures of variations, such as standard deviation and standard error of the mean. If you use data from another published or unpublished source, obtain permission and acknowledge that source fully.

Additional tables containing backup data too extensive to publish in print may be appropriate for publication in the electronic version of the journal, deposited with an archival service, or made available to readers directly by the authors. An appropriate statement should be added to the text to inform readers that this additional information is available and where it is located. Submit such tables for consideration with the paper so that they will be available to the peer reviewers.

i. *Illustrations (Figures):* Digital images of manuscript illustrations should be submitted in a suitable format for print publication. Most submission systems have detailed instructions on the quality of images and check them after manuscript upload. For print submissions, figures should be either professionally drawn and photographed, or submitted as photographicquality digital prints.

For radiological and other clinical and diagnostic images, as well as pictures of pathology specimens or photomicrographs, send high-resolution photographic image files. Before-and-after images should be taken with the same intensity, direction, and color of light. Since blots are used as primary evidence in many scientific articles, editors may require deposition of the original photographs of blots on the journal's website.

Although some journals redraw figures, many do not. Letters, numbers, and symbols on figures should therefore be clear and consistent throughout, and large enough to remain legible when the figure is reduced for publication. Figures should be made as self-explanatory as possible, since many will be used directly in slide presentations. Titles and detailed explanations belong in the legends— not on the illustrations themselves.

Photomicrographs should have internal scale markers. Symbols, arrows, or letters used in photomicrographs should contrast with the background. Explain the internal scale and identify the method of staining in photomicrographs. Figures should be numbered consecutively according to the order in which they have been cited in the text. If a figure has been published previously, acknowledge the original source and submit written permission from the copyright holder to reproduce it. Permission is required irrespective of authorship or publisher except for documents in the public domain.

In the manuscript, legends for illustrations should be on a separate page, with Arabic numerals corresponding to the illustrations. When symbols, arrows, numbers, or letters are used to identify parts of the illustrations, identify and explain each one clearly in the legend.

j. *Units of measurement:* Measurements of length, height, weight, and volume should be reported in metric units (meter, kilogram, or liter) or their decimal multiples. Temperatures should be in degrees celsius. Blood pressures should be in millimeters of mercury, unless other units are specifically required by the journal. Journals vary in the units they use for reporting hematologic, clinical chemistry, and other measurements. Authors must consult the Information for Authors of the particular journal and should report laboratory information in both local and International System of Units (SI). Editors may request that authors add alternative or non-SI units, since SI units are not universally used. Drug concentrations may be reported in either SI or mass units, but the alternative should be provided in parentheses where appropriate.

k. *Abbreviations and symbols:* Use only standard abbreviations; use of nonstandard abbreviations can be confusing to readers. Avoid abbreviations in the title of the manuscript. The spelled-out abbreviation followed by the abbreviation in parenthesis should be used on first mention unless the abbreviation is a standard unit of measurement.

B. Sending the manuscript to the journal

Manuscripts should be accompanied by a cover letter or a completed journal submission form, which should include the following information:

A full statement to the editor about all submissions and previous reports that might be regarded as redundant publication of the same or very similar work: Any such work should be referred to specifically and referenced in the new paper. Copies of such material should be included with the submitted paper to help the editor address the situation. (*See* also Section IIID.2).

A statement of financial or other relationships that might lead to a conflict of interest, if that information is not included in the manuscript itself or in an authors' form. (See also Section IIB).

A statement on authorship: Journals that do not use contribution declarations for all authors may require that the submission letter includes a statement that the manuscript has been read and approved by all the authors, that the requirements for authorship as stated earlier in this document have been met, and that each author believes that the manuscript represents honest work if that information is not provided in another form *see* also Section IIA.

Contact information for the author responsible for communicating with other authors about revisions and final approval of the proofs, if that information is not included in the manuscript itself. The letter or form should inform editors if concerns have been raised (e.g., via institutional and/or regulatory bodies) regarding the conduct of the research or if corrective action has been recommended. The letter or form should give any additional information that may be helpful to the editor, such as the type or format of article in the particular journal that the manuscript represents. If the manuscript has been submitted previously to another journal, it is helpful to include the previous editor's and reviewers' comments with the submitted manuscript, along with the authors' responses to those comments. Editors encourage authors to submit these previous communications. Doing so may expedite the review process and encourages transparency and sharing of expertise.

Many journals provide a presubmission checklist to help the author ensure that all the components of the submission have been included. Some journals also require that authors complete checklists for reports of certain study types (e.g., the CONSORT checklist for reports of randomized controlled trials). Authors should look to see if the journal uses such checklists, and send them with the manuscript if they are requested.

The manuscript must be accompanied by permission to reproduce previously published material, use previously published illustrations, report information about identifiable persons, or to acknowledge people for their contributions.

Appendix 2

Proofreading Symbols

Symbol	Meaning	As typeset and marked for correction		Corrected
ℒ	delete	data/ that we have accumulated	ℒ	data that we have accumulated
ℰ̃	delete and close up	$A(x)\,X\,B(x)$ is the term	ℰ̃	$A(x)B(x)$ is the term
⌒	close up	the product $A(x)\,B(x)$	⌒	the product $A(x)B(x)$
…stet	restore words crossed out	it is not true	stet	it is not true
∧	indicates where to make insertion	colinear	ℓ	collinear
⊙	insert a period ⎫	…in our experiment⊙	⊙	…in our experiment.
⋏	insert a comma ⎬	However₍we …	⋏	However, we…
λ̄	insert a hyphen	un-ionized	λ̄	un-ionized
∧	type or insert as subscript ⎫	$a\sqrt{2}$, A	⌃/⌄	a_2, A^α
∨	type or insert as superscript ⎭			
#	insert a space	1536A	#	1536 A
N̄	en dash	in the range 20–40 MeV	N̄	in the range 20–40 MeV
M̄	em dash	Relation (14)–and only relation (14)⎯can …	M̄	Relation (14) –and only relation (14) –can…
¶	start a new paragraph ⎫	¶The state is represented by the Wheeler form of the vacuum functional.	¶	The state is represented by the Wheeler form of the vacuum functional. Be-
no ¶	do not start a new paragraph ⎭	no ¶Besides the well-known…	no ¶	sides the well-known…
⊔	lower matter ⎫	$a+b =$ ⌐d⌐$+ k$ρρ	⊔/⊓	$a+b = c + k\cdot p$
⊓	raise matter ⎭			
⊏	move matter to left ⎫	$x+y=$ ⌐z⌐$+ w$ (15)	⊏/⊐	$x+y=z+w$ (15)
⊐	move matter to right ⎭			
ℓc	use lower-case letter	liquid-HE container	ℓc	liquid-He container
≡ cap	use capital letter	24.5 meV	cap	24.5 MeV
=sc	use small capital letter	Kr II	sc	Kr II
○ rom	use roman type ⎫	Next I measured I in MeV.	rom/ital	Next I measured I in MeV.
– ital	use italic type ⎭			
∽ tr	transpose	concieve	tr	conceive
∼ bf	make boldface roman	$E \times H$	bf	E × H
≈ bf ital	make boldface italic	$E + H$	bf ital	$E + H$
/	indicates order in which corrections are to be made in a line	parametrizaton	⊙/tr	parametrization

175

Readers' Notes

Readers' Notes

Readers' Notes

Readers' Notes